TECHNIQUES OF STAIRCASE CONSTRUCTION

Technical and Design Instructions
for Stairs Made of Wood, Steel,
Concrete, and Natural Stone

Willibald Mannes

VNR **VAN NOSTRAND REINHOLD COMPANY**

New York

English edition first published in 1986.
Copyright © 1979 by Deutsche Verlags-Anstalt
GmbH, Stuttgart, under the title *Technik des
Treppenbaus*.
Library of Congress Catalog Card Number 86-9252

ISBN 0-442-26086-5

Van Nostrand Reinhold Company Inc.
115 Fifth Avenue
New York, New York 10003

Van Nostrand Reinhold Company Limited
Molly Millars Lane
Wokingham, Berkshire RG11 2PY, England

Van Nostrand Reinhold
480 La Trobe Street
Melbourne, Victoria 3000, Australia

Macmillan of Canada
Division of Canada Publishing Corporation
164 Commander Boulevard
Agincourt, Ontario M1S 3C7, Canada

16 15 14 13 12 11 10 9 8 7 6 5 4 3 2 1

**Library of Congress Cataloging-in-Publication
Data**

Mannes, Willibald, 1925—
   Techniques of staircase construction.

   Translation of: Technik des Treppenbaus.
   Bibliography: p.
   1. Staircase building.   I. Title.
TH5667.M3413   1986   694'.6   86-9252
ISBN 0-442-26086-5

Contents

Foreword

Staircases, which today are equally the responsibility of joiners and carpenters, have had a varied history over the last thirty years. Until 1945 nearly all staircases, even those in large residential blocks, were made of wood. Because of the amount of destruction that took place during the war, new building regulations frequently stipulated nonflammable materials for almost all stairs.

This resulted in a decline in the quality of stair construction; what is more, fewer and fewer craftsmen were trained for this rewarding and varied branch of woodworking craftsmanship. This is a regrettable development, since good stair builders must combine the design capabilities and three-dimensional approach of the carpenter with the exact and neat craftsmanship of the joiner. *Techniques of Staircase Construction* therefore provides welcome guidelines and instructions to those concerned with practical building throughout the construction sequence, from the measuring and the drawing of plans and elevations to the use of materials and the actual construction.

Design engineers and architects will find this book equally useful, since stair construction now consists of two main areas: one comprises ready-made stairs and their varied possibilities, the other, stairs made by craftsmen in a variety of types and styles. In the hope that this second area, the craft-based group, will regain lost ground and that we can speak in the near future not only of a renaissance of wooden structures but also of stairs, I express my good wishes for a wide distribution of this excellent and instructive book.

September 1979

Donat Müller

Chairman of the Federation of German Joiners within the Central Association of the German Building Trade

President of the Associations of Bavarian Joiners and Timber Construction Trades

Introduction

The increasing demand for textbooks on the techniques of stair construction is due to two main factors:

1. The relatively small dwellings that were built twenty to thirty years ago are no longer regarded as acceptable. New regulations concerning noise and heat insulation as well as government aid available to finance such projects have, in addition, stimulated the rebuilding and thus the design of more generously proportioned dwellings, including, of course, staircases.

2. The style of living has changed. The time when sober interiors were the order of the day has gone. Excessive nostalgic reversal to previous styles has also passed.

Natural materials, profitably used by good craftsmen, have once again become fashionable. Turners, ropemakers, potters, and sculptors, whose crafts had become rare, are finding themselves once more in demand. Good craftsmanship is again appreciated and well paid. Many, and above all young, craftsmen have recognized this renaissance. They see a chance to sidestep the competition of mass-produced components by providing tailor-made pieces made to a high standard. Unfortunately, much craft knowledge has been lost. This applies in particular to stair construction, which requires a high degree of expert knowledge, manual dexterity, and design capability.

All those who want to brush up their general knowledge of stair construction techniques or acquire new insights will benefit from this book. It provides basic knowledge for apprentices and students and technical assistance to the designer and producer of stairs.

The first part deals with the general theory of stair construction. It discusses concepts and terminology, dimensions, calculations, and types of construction, all of which are illustrated by numerous drawings and pictures. The most important of the many statutory regulations and recommendations for stair construction are summarized in the second part of this book.

The third part deals with the practical aspects of stairs, especially in domestic buildings. Stair elevations lead to questions of detail for which solutions are required. It is thus possible to elucidate connections that result not only in good construction work but also in the exemplary design of stairs. If wood is given preference over other materials, this is because the qualities of this material have once again been recognized by domestic architects. Moreover, wooden stairs have major design advantages in reconstruction projects.

The fourth and last part of this book gives a number of examples and design ideas.

1. General Theory of Stair Construction

1.1. CONCEPTS, TERMINOLOGY, AND DIMENSIONS OF STAIRS

1.1.1. Plan shapes, including area and space requirements for stairs and staircases

In order to arrive, with usual stair types, at comparable area requirements, the run and rise dimensions of all examples are the same:

Floor height 270.00 = 15 × 18.00 rises (S)
Basic dimension 385.00 = 14 × 27.5 runs (A)

The dimensions for area requirements of stairs (Tr) and staircases (Tr-R) are rounded-off guide dimensions. Depending on the direction in which the rising stairs turn, we refer to right staircases or left staircases.

T = landing depth, B = landing width

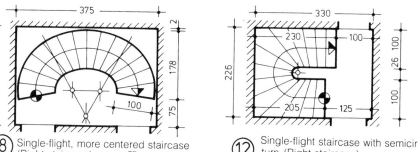

(A) Straight Stairs

① Single-flight straight stairs (Right staircase)
TR = 3.85 sq/m
TR-R = 13.22 sq/m

② Dogleg staircase with half-landing (Right staircase)
TR = 3.85 sq/m
TR-R = 15.23 sq/m

③ Half-turn staircase with landing (Right staircase)
TR = 6.00 sq/m
TR-R = 9.04 sq/m

④ Triple-flight staircase, two turns, with half-landings (Left staircase)
TR = 5.90 sq/m
TR-R = 12.15 sq/m

(B) Curved Stair Flights

⑤ Newel staircase with round well (Single-flight, left staircase)
TR = 3.50 sq/m
TR-R = 6.00 sq/m

⑥ Single-flight, spiral staircase with round well (Right staircase)
TR = 3.50 sq/m
TR-R = 8.12 sq/m

⑦ Single-flight, semicircular staircase (Right staircase)
TR = 3.90 sq/m
TR-R = 10.50 sq/m

⑧ Single-flight, more centered staircase (Right staircase)
TR = 4.50 sq/m
TR-R = 9.56 sq/m

(C) Stair flights Consisting of Straight and Circular Components

⑨ Single-flight staircase with quarter turn at the starting section (Right staircase)
TR = 2.90 sq/m
TR-R = 11.75 sq/m

⑩ Single-flight staircase with quarter turn at the top section (Right staircase)
TR = 2.85 sq/m
TR-R = 11.75 sq/m

⑪ Single-flight staircase with two quarter turns (Left staircase)
TR = 2.50 sq/m
TR-R = 9.00 sq/m

⑫ Single-flight staircase with semicircular turn (Right staircase)
TR = 4.60 sq/m
TR-R = 7.46 sq/m

1.1.2. Dimensional terms relating to stairs

Designation of stair and railing parts

Railing Parts Made from Wood

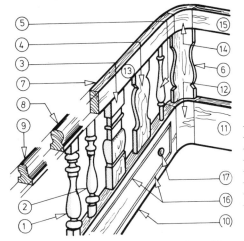

1. Bannister, middle section turned, square ends
2. Bannister, turned
3. Rectangular bannister
4. Flat bannister with shaped edges
5. Turned handrail standard
6. Rounded bannister
7. Rectangular handrail
8. Symmetrically shaped handrail
9. Overhanging handrail
10. Face string
11. Blind string
12. String wreath
13. Handrail
14. Handrail wreath
15. Handrail
16. Ornamental beading
17. Ornamental rosette

Dimensional Terms Relating to Stairs

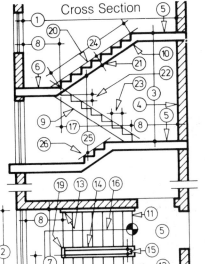

Cross Section

Plan

1. Stairwell, length
2. Stairwell, width
3. Floor height
4. Stairwell wall
5. Landing
6. Half-landing
7. Landing width
8. Landing depth
9. First flight
10. Top flight
11. First step
12. Top step
13. Wall string
14. Face string
15. Stairwell
16. Walking line
17. Flight length (basic length)
18. Flight width
19. Landing (turning) post
20. Stair pitch
21. Stair depth
22. Rise (S)
23. Run (A)
24. Flight length over pitch
25. Headroom
26. Differential staircase height

Railing with Wooden Paneling

1. Molded handrail
2. Flute
3. Molded beading
4. Shaped panel
5. Profiled paneling
6. Chamfered beading
7. Rounded beading
8. Fluted beading
9. Ovolo
10. Cornice

Handrail Parts Made from Steel

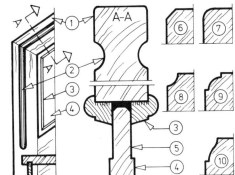

1. First handrail standard (main standard), hand-forged
2. Bar, turned
3. Openwork twist
4. Base (cover) rosette
5. Forged handrail with curved end section
6. Lower chord
7. First standard (square standard)
8. Bar (flat steel)
9. Upper chord
10. Lower chord
11. PVC handrail

Designation of Stair Parts

1. Wall string
2. Face string
3. Section width of string
4. Step (tread)
5. Riser
6. Undercut
7. Height of rise (S)
8. Width of tread (A)
9. Basic dimension of staircase
10. Floor height
11. Solid rectangular step
12. Stair bolt
13. Handrail standard, first step
14. Handrail standard, top step
15. Handrail
16. Handrail standard (cylindrical)
17. Handrail standard with rosette
18. Blind string
19. Balustrade handrail
20. Ceiling height, unplastered
21. Finished height, top
22. Finished height, underside
23. Overhang blind string

Handrail with Glass Paneling

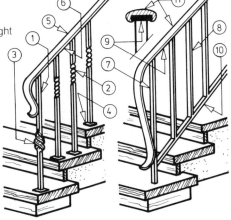

1. Glass (safety glass, acrylic glass, wire glass)
2. Edges ground (polished)
3. Fitted in groove
4. Clamped
5. Holding plates
6. Standard
7. Handrail

1.2. DESIGN TYPES OF STAIRS AND STAIR RAILINGS

1.2.1. Stairs made from wood

Saddle-type stairs with supporting strings and brackets

Saddle-type stairs with doweled wooden brackets. The tread is screwed on from the underside of the bracket; subsequently, tread and bracket are dowled onto the supporting string.

Saddle-type stairs with twin supporting strings. A modern rustic construction that can, in a nailed version, be used for temporary staircases.

Saddle-type stairs with tread supports. Round wooden or steel supports, about 30–35 mm thick, are placed below the treads, which are supported at four points and should be cross-banded.

Saddle-type stairs fastened with steel angles. A strong connection between supporting string and treads.

Saddle-type stairs with handrail standard. The supports are notched; the underside of the steps is fastened at the back by dowels. Handrail standards and ends of treads are notched, doweled, and glued together.

Saddle-type stairs with solid block steps. Steps and supporting strings consist of glue-laminated layers and are linked by strong dowels. This design is particularly suitable for large staircases.

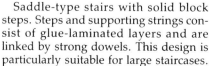

Saddle-type stairs with notched supporting strings.

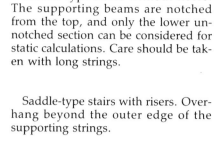

Saddle-type stairs without risers. The supporting beams are notched from the top, and only the lower un-notched section can be considered for static calculations. Care should be taken with long strings.

Saddle-type stairs with risers. Over-hang beyond the outer edge of the supporting strings.

(a) Risers approximately 1–3 cm
(b) Other steps approximately 4–6 cm

Saddle-type stairs with box-shaped supports. The supporting element consists of two beams made up from glued planks and core board. (The hollow section should be filled with glass wool or other noise-insulating materials.)

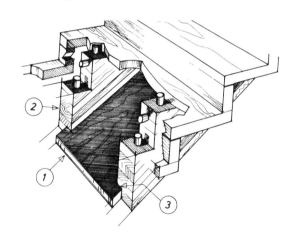

Closed string staircase.

Mortised stairs. Risers and treads are mortised to a depth of about 2 cm into 5 to 6-cm-thick strings. The step is held together by continuous bolts, about 10 mm in diameter.

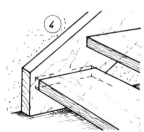

Mortised open-riser stairs. Mortised stairs without risers. The visible bolts may be sunk into the treads.

Design variations of mortised stairs.

1. Stairs with tenons and wedges
2. Treads additionally notched
3. Treads notched, corners cut off and bolted onto the string.

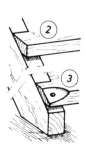

1.2.2. Stairs made predominantly from steel

1. Saddle-type stairs. Rectangular steel tubing used for supporting beams (1). Flat steel bent accordionlike in the shape of stairs is bolted or welded onto the string as support for the treads. These can be made from cast stone or wood and are fastened from underneath. Felt or similar material should be placed between treads and steel. The supporting string should be filled with foam for maximum noise insulation when the stairs are in use.

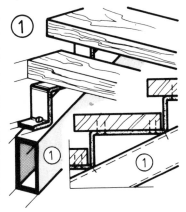

2. String staircases. Strings made from rectangular steel tubing with welded-on angles supporting the treads.

3. String staircases. Risers and treads are formed by bending steel sheet. This is welded onto strings made from steel channels (1). Textile covers can also be fitted to the underside of the stairs. The strings may also consist of wood, into which the steel sheet can be slotted and bolted (2).

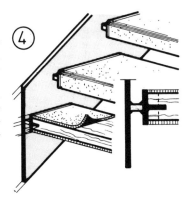

4. String staircases. T-shaped sections are welded onto the approximately 15-mm-thick steel sheet string. The steps are grooved and screwed onto the brackets. The tread thickness should be reduced all around by the thickness of the textile cover.

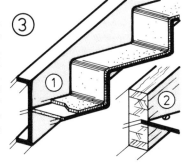

5. Stairs supported by a tubular steel frame. Instead of a face string, a welded tubular steel frame made from rectangular tubing supports the treads. The frame is fitted between the ceilings of the floors or extends over a length of about 90 cm over the upper floor and serves at the same time as balustrade. The treads are bolted on from underneath.

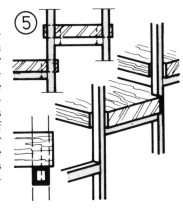

1.2.3. Stairs made predominantly from stone and concrete

1. Stairs attached on one side only. Treads inserted into recesses of the wall and linked on the other side by a bolt. (In order to provide noise insulation for adjacent rooms, rocker-type anchors should be used as supports.)

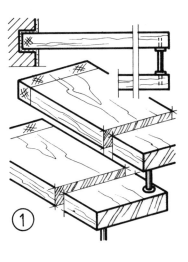

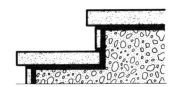

2. Risers and treads made from cast or natural stone placed onto plain concrete and embedded in mortar. The steps should project laterally by about 4–5 cm, with an overhang of the riser of 2–3 cm.

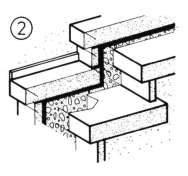

2(a). In this example, only treads but no risers are placed onto the substrate.

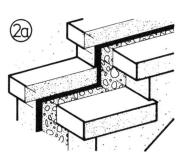

Stairs with cantilever treads. Fair-faced concrete slabs (1) are fitted into a 24-cm-thick wall. Channel-shaped glue-laminated board (2) and carpeting (3) attached from the underside. The wooden board is attached to the concrete slab by special anchoring devices (4).

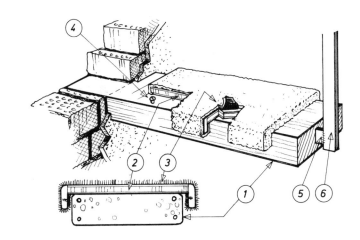

1.3. DESIGN TYPES OF STAIR RAILINGS

1.3.1. Railings made from various materials

The main function of stair railings is to provide safety. The intervals between standards and the height of the handrail are determined by existing building regulations (see also page 21). Bannisters can have round, square, or rectangular cross sections of varying sizes.

Stair railings with bannisters

1. Stair railings with L-shaped handrail and cylindrical bannisters made from wood or metal. Small-diameter bannisters should be fitted to a depth of 6–7 cm when they act as restraints.

2. Stair railings with T-shaped handrail, cylindrical bannisters, and rectangular intermediate member. The installation of an intermediate member provides thin bannisters with greater stability.

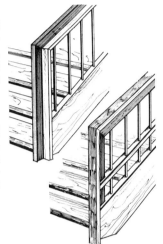

3. Stair railings with oval handrail and fishbelly-shaped bannisters. This type of handrailing, which has been used for many years, can provide an interesting design feature.

4. Stair railings with panels. Safety glass or acrylic glass panels fitted between handrail and string. As can be seen from the sketch, stairs and railings form a design unit. In order to avoid unattractive covering strips, the panel has to be incorporated during the actual stair/railing construction. The curved corners are highly labor intensive but provide a distinctive feature.

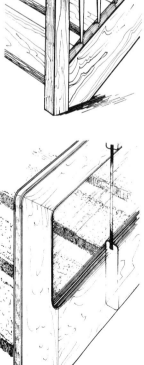

5. Stair railings with wooden or metal bannisters attached to the outside of the handrail. No paneling. A clear modern design that permits many variations. The supporting rods can be attached either by invisible dowels or by visible screws.

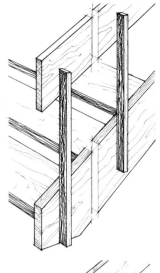

6. Handrail with steel rods outside the handrail and midrail sideways onto the stairs.

Cross section of steel rod, sideways onto the stairs, approximately 10 × 40 mm.
Cross section of the steel rod, lengthwise to the stairs, approximately 15 × 40 mm.
Cross section of the steel rod, square to the stairs, approximately 22 × 22 mm.

The midrail should be at least 22 mm thick. Bolting of strings, handrail, and midrail is from the stair side. Stair railings using these support rods require no standards. Handrail and midrail can either be made as an integrated frame or as individual units.

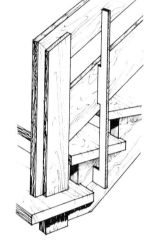

7. Stair railing with steel supports fastened on the outside. These railings use an infill made from flat steel rods. This gives an impression of lightness, and such constructions should preferably be used in narrow staircases. The stability of the stair railing in the direction of the stair flight is less than that provided by a midrail; the standard at the bottom step therefore has to withstand the forces acting in this direction.

8. Saddle-type treads notched to fit bannister boards. The treads, into which the bannisters are fitted, are equipped with special fastening sleeves on the face side, since the thread of wooden screws would not grip properly in end-grain blocks. 20-mm wooden dowels may be glued into the holes above the fastening devices: their ends may project between 5 and 10 mm above the surface of the board.

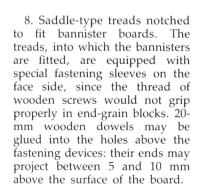

9. Stair railings with rope in lieu of a solid handrail. When a load is placed on the steps, the direction of the line of force is toward the outside; through the bannister rods it returns, on the step below, toward the center and the notched supporting beam. The tops of the bannister rods are bent so that the rope can be threaded through.

12. Wrought iron stair railing. Starting post with openwork twist; the ornamental ironwork has been hammered flat. The individual units are joined by brass sleeves; the forged bottom chord is bent to match the staircase and the handrail shaped at the edge, its top hammered.

13. Stair railings consisting of vertical boards. This rustic design, which permits many variations, can be fitted to newel staircases. It can also be made into an attractive feature of string staircases, with the treads projecting through the board.

10. Stair railing with midrail and steel-tube handrailing. Box profile midrail, with the gap equaling the thickness of the vertical rods. This substantial design is particularly effective when painted.

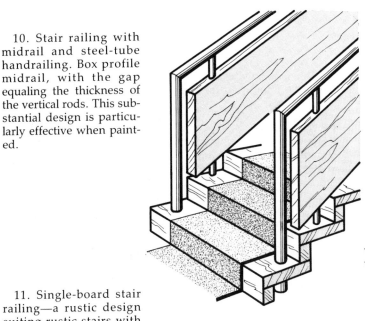

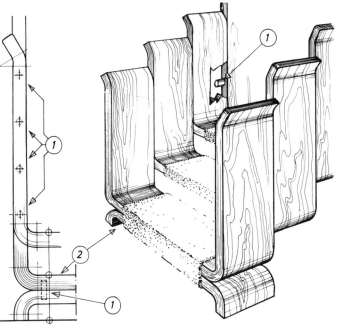

14. Veneered molded staircases. Stairs consisting of cores onto which veneers have been fitted and glued. Dowels must be used when bonding vertical cross joints.

11. Single-board stair railing—a rustic design suiting rustic stairs with treads slotted into position. The wood is solid throughout. All fastenings are visible as a special feature.

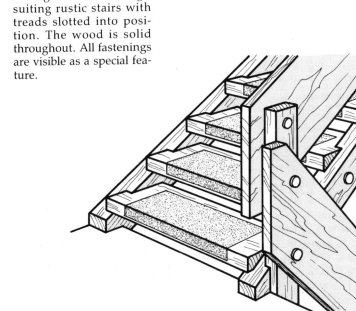

1.3.2. Railings with load-bearing handrail (stairs with suspended treads)

In the following examples, the handrail forms part of the static stair design and fulfills a load-bearing function. By skillfully utilizing these elements, it is possible to build stairs at a reasonable cost that nevertheless give the impression of lightness. It is advantageous to anchor the handrail of a straight flight of stairs where it intersects the ceiling. This greatly reduces the effective span. The rods have a safety function and act as suspension or tension rods.

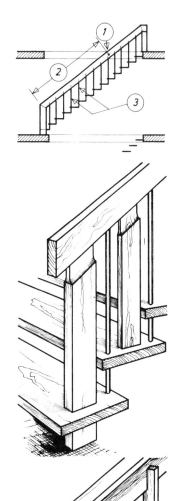

Stairs with suspended steps and round steel tension rods. In the sketch shown above, the standard supports the lower part of the handrail. Starting from the handrail, tension rods pass through the treads that are to be connected to each other. Tension rods can consist of 20–22 mm steel tubing that passes through the treads and is bolted onto these horizontally. In the case of stairs subject to much traffic, they may be made from threaded round bars. The board inserted between the bannister rods is a safety feature.

Stairs with suspended treads and tension rods made from hard textured wood. The suspension rods shown in the middle picture do not provide maximum stability. The structure is held together at the top by a strong dowel. It is important to provide a sufficient length of wood between the dowel and the end of the rod so that no splitting can occur when loads are applied. The cross section of the handrail is in the form of a T or L so that the upper pressure zone of the handrail is protected against lateral buckling.

Stairs with suspended treads and flat steel tension and cross members. Strong welded construction consisting of flat bars (approximately 10 × 40 mm) and attached laterally to the handrail.

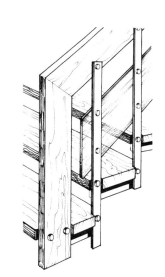

1.3.3. Handrail profiles

1.3.3.1. Handrails made from wood

Today, most handrails are made in the general shape of a rectangular plank. They are nonslippery so that they provide support and security, especially to older people. The shape of the handrail should match the style of the staircase, and its cross section should be in proportion to its span (50 kg/m², horizontal pressure). The type of profile chosen for straight or circular handrails has a significant bearing on calculations. All handrails should be made from hard textured wood and given a good surface finish, so that they do not look greasy after a short period of use.

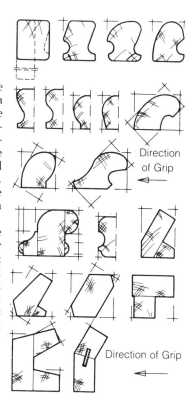

1.3.3.2. Handrails made from steel, nonferrous metals, or plastics

Handrails made from pure steel appear fresh, while naturally treated wooden handrails on steel stair railings provide warmth and a feeling of comfort, and metal handrails attached to predominantly wrought iron handrails give greater elegance. Special handrails are the prerogative of large stair constructions. Where the handrails are relatively slim, the standards should be arranged at an angle, so that fingers can slide past them.

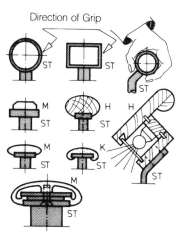

Enlarged

ST = Steel
M = Metal (brass, bronze)
H = Wood
K = Plastic

1.3.3.3. Grab rails

The distance between grab rails and wall should be at least 4 cm, the depth 8 cm. Lights installed behind these handrails are an architecturally interesting feature and provide additional security. Note, however, that if the walls are badly rendered they appear cloudy.

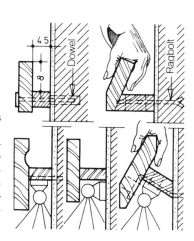

1.4. DESCRIPTION OF STAIRS AND DESIGNATION OF STAIR PARTS IN COMPLETED PROJECTS

1. Half-turn stair with landing and a centrally arranged "spine" string made from in-situ cast concrete. The treads are designed as solid rectangular steps, laid in cement mortar. Railing: flat steel with wooden panels, handrail made from oak wood, fitted into a steel chord.

2. Single-flight, straight, saddle-type staircase; the centrally positioned spine string is made from glue-laminated pinewood. The treads are made from cross-banded oak and are glued and doweled with about ten dowels onto the notched beam. The stair railing is in the form of a glue-laminated pine board, with tubular steel supports forming the connection between treads and railing.

3. Single-flight, spiral, semicircular, saddle-type staircase; the centrally positioned spine string consists of in-situ concrete, the treads of marble. The treads are anchored on the outer edges by chromium-plated steel bolts and attached to the spine string by metal ties and concrete mortar. Stair railing and balustrade are made from chromium-plated tubular steel with mounting devices for the acrylic glass panels. Handrail: glue-laminated oak.

4. Straight saddle-type staircase. The two oak support strings are glue-laminated and fitted with doweled brackets. Stairs: solid oak. The stair railing consists of cylindrical rods with a rectangular handrail of solid oak, which is fitted into and doweled onto the first and last standards. Bannister rods: ash.

5. Single-flight semicircular staircase, saddle-type. The supporting strings consist of rectangular steel tubing, welded together in the shape of steps; the treads are oak, bolted onto the supports from underneath. The bannister rods of this wrought iron stair railing are either twisted or decorated with openwork twist, while the main standards at the top and bottom are twisted.

6. Half-turn saddle-type staircase with landing, string beams from rectangular steel tubing, with the angular intersection between stair flight and landing being welded. The treads are supported by flat, angular steel welded onto the beam. The treads, consisting of laminated chipboard (V-100) are glued together in several layers, with the bottom layer and edge of the back consisting of 1.5-mm teak veneer. Ends: solid teak; top and front edge: textile cover; bannister rods: chromium-plated bright steel.

7. The saddle-type treads and risers of this staircase are marble. The edges of the treads project by about 5 cm, those of the risers by about 3 cm from the concrete core. The handrailing consists of steel rods with panels of safety glass, with the tubular steel rods and welded-on channels supporting the glass panels. Handrail: solid mahogany.

8. Single-flight semicircular staircase, mortised, open riser. Wall string: solid red fir; face string: 3-mm fir veneer, glue-laminated. Steps: solid oak, cured. Handrail: glue-laminated oak, slightly molded. Bannisters: bent and glue-laminated boards, shaped at the edges.

9. Single-flight semicircular staircase, mortised, open riser. The four large stair segments are suspended from the ceiling by V-2 steel rods (suspended staircase). Stringers, standards, and handrails are, at the respective joints, tied together by strong steel ties and dowels. The steps consist of cross-banded laminated core board, veneered on top with a 6-mm sawn veneer, on all other sides with a 3-mm sliced oak veneer.

10. Single-flight semicircular spiral staircase, mortised, open riser. The strings consist of glue-laminated pine veneer. The blind string elements are fitted with glass panels, consisting of 10-mm security glass. Steps: solid oak, cured. The wall string should be anchored centrally in order to avoid a pendulum effect.

11. Single-flight semicircular stairs, mortised, open riser. The wall string consists of solid oak, the central string of a glue-laminated solid oak block. The steps are of laminated wood (vertical glue joints) with a textile cover. On the wall side, wooden handrails and on the inside of the stair railing, a 35-mm-diameter rope attached to browned metal loops.

12. (Left) Spiral staircase, single-flight, stylized, mortised, open riser. Strings are glue-laminated, the face string decorated with ornamental beading and covering rosettes. The bannister rods are turned wood, the handrail standard at the bottom step has molding all around.

12. (Right) Single-flight spiral staircase, mortised, stylized. Bannisters in the form of turned rods. Turned handrail standard at the bottom of the step with a helically molded end of the handrail.

13. Half-turn staircase with landing (quarter-landing), mortised face string with string wreath. The wrought steel stair railing consists of bent, forged flat steel, joined by brass sleeves. Lower chord from flat steel, upper chord flat brass profile.

14. Single-flight stairs, slightly curved at the bottom step, mortised, stylized. The bottom step is shown by the plan to be a slightly rounded solid rectangular step; the starting post is rectangular and carved on both sides; bannister rods turned (diameter more than approx. 50 mm) with square-edged top and bottom sections.

15. Double-flight staircase with quarter-landing, closed string, stylized. Face string with quarter string wreath. The handrailing is designed as a panel structure with ornamental beading. All components shown by the plan to be circular consist of glue-laminated oak veneer. The blind string of the balustrade covers the edge of the floor.

16. Single-flight straight staircase, mortised, treads and risers made from core-wood, top and bottom surfaces with textile cover. The wall string is distant from the wall by about 8 cm. The face string with top and bottom handrail standards and the handrail consists of various layers of overlapping core wood, glued together to form one unit. 10-mm acrylic glass panels are fitted all around into slots; wall-string and face-string elements have a plastic coating.

17. Half-turn staircase with half-landing, mortised. The glue-laminated pine strings are anchored to the reinforced concrete landing slab. The face strings are joined at the turning point by connecting pieces. The stair railing consists of bannister rods and a rectangular handrail with mitered corners at the turning point. Bannister rods: 18-mm bright steel, browned.

18. Single-flight straight staircase, mortised into the wall string, whose outline follows the line of the stairs. On the face side, steps suspended by steel rods. The handrail has a support function and must be secured against lateral buckling. The handrail standard at the bottom step is anchored to the first step by a strong steel angle piece.

19. Suspended steps, cross-banded oak, mortised on the wall side; tread/string anchoring by two lock nuts, M8. For this purpose, 40-mm-long sleeves should be fitted into the face of the steps. On the face side, treads suspended by board-type bannisters, shaped at the edges. Connection between steps and bannisters by wedges.

20. Single-flight spiral staircase. Supporting structure from in-situ cast concrete below exposed concrete, top surface with textile cover. The glue-laminated wall and face strings must be fitted before the textile cover, working from the top surface of the steps. Bannisters and handrails are glue-laminated. The 10-mm acrylic glass panels should be dimensioned as large as possible. They are fitted into grooves in handrail and strings and secured by adhesive.

21. Newel staircase, single flight. Newel consisting of several glue-laminated planks. The pine treads have been slotted into the newel to a depth of about 3 cm and are secured by hardwood dowels passed through the newel. Fastening on the wall side by flat metal strips fitted into the steps.

22. Newel staircase, single run. Newel consisting of seamless steel tube, with T-shaped sections, welded onto the tube, supporting the steps. Treads consist of laminated 13-mm corewood, with a textile cover, the face being formed by curved, glue-laminated hardwood. Handrail: glue-laminated beechwood. The handrail standard at the bottom step passes through the step and is secured underneath to the horizontal T section by welding. Circular steel rods stabilize and support steps and handrail.

23a. Single-flight newel staircase, steel structure. The newel post consists of seamless tubing, the steps of 3-mm steel sheeting attached to the 8-mm handrail standards by welding to form a basic unit. Polished V-2a steel (Nyrosta) was chosen for the handrail. After installation, the inner section of the steps should be painted with a noise-insulating compound, filled with concrete up to 7 mm below the top edge of the steel sheeting, and covered with a textile cover.

23b. The balustrade of the newel staircase described above (23a) corresponds with the construction of the stair railing. The stairwell opening in the ceiling is fitted with a circular strip of steel sheeting up to the level of the finished floor. Here, too, the screed is about 7 mm below the top edge, so that the facing of the carpet is covered.

1.4.1. Space-saving stairs

1.4.1.1. Space-saving straight stairs

Steep stairs of a standard tread design are difficult to negotiate, especially in descent, when the leg will feel its way along the front edge of the second step (1), as shown in the illustration, and push as far back on the first step (2) as possible. The ball of the foot remains unsupported (3). This position is very hazardous, but the problem can be resolved if the step that is being passed is narrower (4), so that its front edge is no longer in the way of the descending leg. This means, however, that the stairs must always be negotiated with the same foot first.

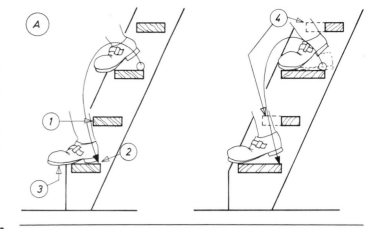

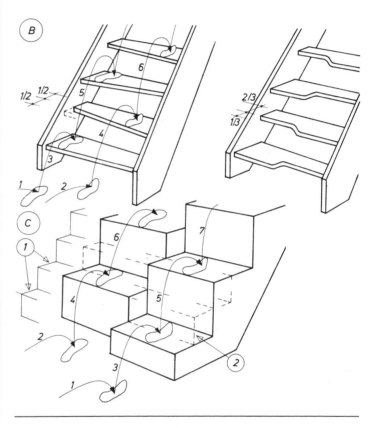

1.4.1.2. Space-saving spiral stairs

This type of stair is ideally suited for spiral staircases with an internal clear passage of less than 70 cm. With treads shaped so that they are deeper on the side of the newel, such staircases provide maximum safety.

The circular design shown in our example virtually forces one to start with the right foot, especially as the top landing is cut so as to create a narrow tread on the side of the newel.

1.5. TYPES OF STEPS

1.5.1. Types of steps, defined by the cross section of the step

(A) Solid rectangular steps
Steps with rectangular or nearly rectangular cross section (solid or with cavity).

1. Solid rectangular step; because of horizontal movement, restricted use for certain tread widths.

2. Solid rectangular step with rabbet (1) (prevents movement toward the front).

3. Solid rectangular step with beveled undercut (1).

4. Solid rectangular step with rabbet (1) and rounded molding at the back (2) (very difficult to manufacture).

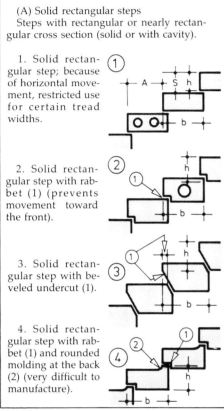

(B) Spandrel steps (triangular steps)
Steps with triangular or near-triangular cross section (solid or with cavity).

5. Spandrel steps with broken rabbet (1) (prevents movement toward the front).

6. Spandrel step, universally suitable for stairs with a pitch between 30° and 45° (solid or with cavity).

7. Spandrel step, Finnish standard step, suitable for any standard plan length or floor height.

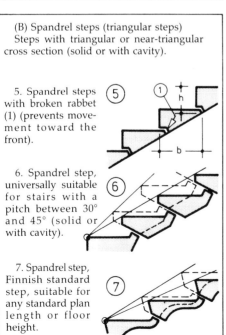

(C) Slab steps
Steps with rectangular or near-rectangular cross section.

8. Slab step suitable for all riser heights and tread widths.

9. Slab step, reinforced along the underside of the front edge.

10. Slab step, with reinforcement at the top of the rear edge.

(D) Angle-type steps
Steps with angular (L-shaped) cross section.

11. Angle-type step ground on the outside (1).

12. Angle-type step ground on the inside (1) (very time-consuming to produce).

13. Angle-type step ground on the outside (1) with beveled undercut (2).

14. Angle-type step interlocking steel sheeting (predominantly for exterior stairs).

(E) Dimensioning of steps
Designation of step faces
Edge finishes

Front Edge

Front Edge of Step

Dimensioning
1 = Length of step
b = Width of step
h = Height of step
d = Thickness of step

Step surfaces
1 = Tread/top surface
2 = Bottom surface
3 = Front edge
4 = Rear edge
5 = End facing

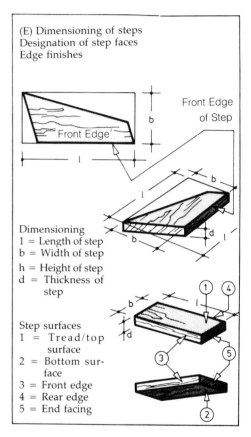

(F) Risers
According to their function, these are not steps but are installed vertically as filler slabs between treads.

1. Riser between treads (1).

2. Riser recessed (undercut) (2).

3. Riser (1), preferred construction for wooden stairs (fitted into grooves of slab steps) (2).

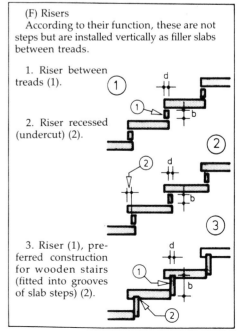

18

1.5.2. Types of steps, defined by materials

Exterior steps should be frost- and acid-resistant (for example, granite). It is not possible to go into details of stair coverings made from ceramic or other tiles. Certain types of exterior stone steps require official permission. The surfaces should be slip resistant. Oiling or waxing is not recommended.

Stone Steps

(A) Solid stone steps made from solid natural stone (marble), sawn.

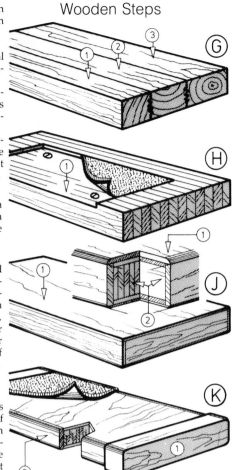

(B) Concrete steps with steel reinforcement. Grain size 0–20, 0–30.

(C) Cored steps; core consisting of reinforced concrete, outside of natural stone.

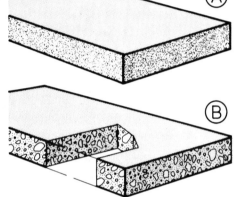

(D) Laminated marble steps; marble slabs bonded together with a special adhesive.

(E) Plate-type marble steps, bonded with concrete; steel reinforcement with large marble slabs and marble corner pieces.

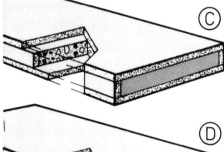

(F) Steel sheet trough-type steps with steel reinforcement, filled with concrete (staircases suitable for construction purposes or for permanent use if subsequently covered with carpet).

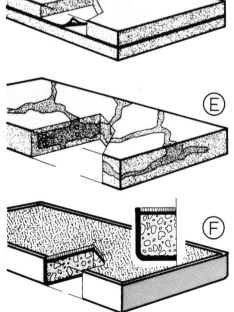

Wooden Steps

(G) Solid wooden steps, glue-bonded in several places.

1. Wood with vertical annual rings (highly recommended)
2. Wood with horizontal annual rings (less recommended)
3. Heartwood. Heartwood must not be used; the heart should be cut out.

(H) Steps made from laminated boards, with cutout (1) for textile cover.

(J) Cross-banded steps; tread (1) to consist of high-grade timber, not less than 5 mm thick, edges 2.7 mm, bottom surface veneer of about 1.5 mm. Inner core (2) consisting of chipboard or coreboard.

(K) Plywood steps with solid end (1) (if steps are covered with textiles, rounded hardwood edges should be glued onto the front edges).

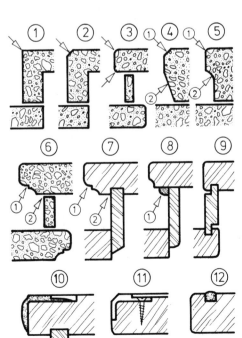

1.5.3. Surface finishes

Stone: Quarry face, sawn, lightly ground, ground, polished, treated with fluosilicate.

Wood: Rough-sawn, rough-machined, planed, ground, stained, cured, matted, varnished, sealed.

1.5.4. Edge finishes

1. Angle with sharp corners
2. Beveled
3. Rounded
4. Beveled (2) front edge
5. Rounded (1) with round-molded front edge (2)
6. Molded (1) with recessed riser (2)
7. Molded (1) with riser fitted into groove (2)
8. Molded with reinforcing bar (1)
9. With discontinuous front edge and flush riser
10. PVC stair nosing
11. Metal stair nosing
12. PVC strip toughened with diamond powder

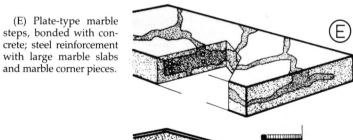

2. Regulations and Recommendations

2.1. STATUTORY REGULATIONS

Building regulations issued by the various regions of the Federal Republic of Germany, which may differ from one another and should be standardized.

2.1.1. Regulations on Implementation

Executory regulations regarding the regional buildings regulations
Regulation of the trade inspectorate
Regulations of local accident underwriters
Regulations on places of assembly
Regulations on stores and business premises
Special regional regulations on public buildings.

2.2. DIN STANDARDS

DIN 18064 Part 1, Stairs:
Concepts, terminology

DIN 18065 Part 1, Stairs in domestic buildings: main dimensions
DIN 4174 Floor heights and stair rises
DIN 18334 Carpenters' and wood construction work
DIN 68360 Wood for carpentry work, quality standards
DIN 68365 Timber for carpenters' work, quality standards
DIN 68368 Planks for stair construction, quality standards
DIN 68705 Part 3, Plywood, building board, veneer board, quality standards
DIN 68705 Part 4, Plywood, building board, veneer board, quality standards
DIN 68763, Chipboard, pressed sheeting for building purposes, designations, properties, examination, monitoring
DIN 1052 Part 1, Timber structures, design and execution
DIN 1055 Part 3, Design loads for structures, live loads
DIN 4074 Part 1, Timber for wooden structures, quality regulations for cut timber (conifers)
DIN 4102, Behavior under fire of building materials and components
DIN 4174 and 18064 are fully quoted in the Appendix to this book
DIN 18065 is at present being revised. Regulations from this DIN standard to which reference is made in this book may be subject to revision.

2.3. IMPORTANT EXTRACTS FROM DIN STANDARDS

Structural stability:
According to DIN 1055, Part 3, Design loads for buildings; live loads, the following live loads should be assumed:

In domestic buildings 3.5 kN/m^2
In public buildings 5.0 kN/m^2

Treads without risers should be designed for a single load in the least favorable position:

In domestic buildings   1.5 kN
In public buildings   2.0 kN

At the level of the handrails, the following horizontal loads should be assumed:

In domestic buildings   0.5 kN/m
In public buildings   1.0 kN/m

Fire Protection
According to all regional building regulations, stairs in buildings with up to two full stories and a total area of 500 m^2 may be made from flammable materials. In the Federal provinces of Bavaria and Schleswig-Holstein, stairs in buildings with three to five full stories and a floor area of more than 500 m^2 must be made so that they are fire retardant. In the case of wooden staircases, this can be achieved by adhering to minimum cross sections in accordance with DIN 4102 or by claddings. In the other Federal provinces, regulations require that in buildings with three or more full stories, the stairs must be made from nonflammable materials. In Baden-Würtemberg, the reguation reads as follows: "The load-bearing sections of necessary staircases should be made from hard wood or nonflammable materials in the case of buildings.

1.   With three full stories, where the floor area of the building, divided into fire lobbies, does not amount to more than 500 m^2 per fire lobby.
2.   With four or five full stories.

If the requirement is for a flame-retardant (F30) design, DIN 4102, Part 4, applies to design and execution."

Rise/Run Ratio
The permissible rise/run ratio of a staircase in a domestic building is determined by DIN 18065, Sheet 1. According to this, the rise in a single family unit must not exceed 21 cm; that in a multi-family building with two dwellings must not exceed 19 cm. The minimum depth of tread in a single-family unit is 21 cm; that in buildings with two dwellings and more, 26 cm. If the tread dimension is less than 26 cm, an undercut of at least 3 cm must be provided (i.e., the tread must project from the riser by at least 3 cm). In the case of high stories, tread dimensions of as much as 29 cm are required.

Stair Railings
According to the regulations on the implementation of prototype statutory requirements, the following applies (with the exception of some regional regulations):

Stair railings must be at least 90 cm high, measured from the front edge of the step, or 1.10 m high in the case of stairs with a drop of more than 12 m. In the case of newel staircases, stair railings on the inside may have to be as high as 1.10 m.
Other necessary restraints must have the following minimum dimensions:

1. Restraints safeguarding openings in ceilings and roofs, subject to foot traffic, and restraints around areas with a drop of up to 12 m: 0.90 m
2. Restraints around areas with a drop of more than 12 m or balustrades in walkways leading to safety staircases: 1.10 m

In buildings where the presence of children must be assumed, apertures in necessary handrails, balustrades, and other restraints must not exceed 12 cm if the drop is more than 1.50 m.

Horizontal gaps between the restraint and the area to be safeguarded must not exceed 4 cm. Restraints installed in accordance with point 1 above should be designed so that they cannot be easily climbed by children.

In addition, the executory orders of the individual provinces apply.

Design of Steps

According to DIN 18065, an undercut of at least 30 mm is required if the tread depth measures less than 260 mm. Certain building regulations require, in the case of stairs without risers or where the presence of children must be assumed, that the clear gap between steps must not exceed 120 mm.

Headroom

The headroom is defined as the distance between the front edge of the step and the underside of the flight of stairs or of other building parts above, measured at right angles. According to the prototype building regulations, this dimension should be at least 2.00 m. This distance is insufficient, since one is inclined to lean forward when descending stairs. According to DIN 18065, the headroom should be at least 2.10 m.

Staircase Width

According to the prototype building regulations, the following utilizable width should be provided (with few exceptions, these figures apply to the whole of the Federal Republic):

	Stair width
1. Stairs in single family units	0.80 m
2. Stairs in domestic dwellings with up to two full stories	0.90 m
3. Stairs in domestic buildings with more than two full stories and stairs in other buildings	1.00 m
4. Stairs leading to basements and attics, emergency stairs	0.80 m
Exceptions: Berlin	
No exceptions for single family units, narrowest flight of stairs	0.90 m
Berlin and Hamburg: Stairs in domestic buildings with more than two full stories	1.10 m

For stairs used by more than 125 persons, greater widths may be required. In the case of stairs that are infrequently used, in particular stairs not leading to living rooms, reduced width may be accepted. The utilizable width is measured at handrail level between the surface of the wall and the inner edge of the handrail or between two handrails.

Noise Insulation

Transfer of sound to adjacent living rooms.

Where staircases are next to living rooms stair components should, where possible, not be tied into the separating wall; otherwise, noise-insulating measures should be taken.

Traffic noise may be reduced by using textile or PVC coverings.

In the case of mortised wooden steps, noise can be reduced if noise-insulating materials (mineral wool, polystyrene) are fitted at the back between tread and riser and the cladding on the underside of the stair flight.

In the case of mortised steps without risers, thin treads may bend and creak in the contact area with their supports.

Safety in Dark Areas

Since the hazard of falling is increased on dark staircases, light-colored steps should be used or the front edge of steps marked by light stair nosings. Stair lights must not dazzle. Stairs with fewer than three risers or individual steps can easily be missed and are a hazard. Stairs without risers allow light to shine through and appear lighter.

Slip-resistant Treads

Oiled and waxed surfaces represent a great hazard and are not permissible. PVC coverings should be kept clean with cleaning agents recommended by the manufacturer. Sealed surfaces should simply be wiped over with a damp cloth and lukewarm water. Aggressive or slippery cleaning agents must never be used, since they normally preclude the subsequent use of sealants.

3. Calculation and Design of Elevations for Stairs and Stair Parts

3.1. CALCULATION OF STAIRS

3.1.1. Stair pitch

Measured on plan, the normal stride of a medium-sized person is between 70 and 75 cm long. This distance becomes shorter when one walks up or down an incline. Inclined surfaces (ramps) are relatively easy to negotiate up to an angle of about 15°. In this pitch range it is recommended, especially in gardens or other outdoor areas, where ice may occur, to install steps with horizontal or inclined treads.

From 20° upwards, the slope should definitely be broken up into steps with horizontal treads. As a result, varying rise heights (S) and tread depths (A) arise. Several adjacent steps form a flight of stairs. The pitch of stairs varies between 20° and about 75°, that of ladders between 75° and 90°, that of ramps between 0° and about 20°.

The ideal pitch for a domestic staircase is about 30°. This also provides the ideal rise/run ratio.

3.1.2. Rise/run ratio

In order to make stairs comfortable to negotiate, the ratio of tread depth and rise height should be correct and follow the so-called rise/run ratio.

The most frequently used rule in the design of stairs states that the sum of two rises (S) and one tread (A) should be between 60 and 65 cm and be as close as possible to 63 cm, as shown in the following formula:

$$2 \times (S) + (A) = \text{approx. } 63 \text{ cm}$$

The ideal rise/run ratio for domestic stairs is:

$$2 \times (S) \ 17 \text{ cm} = 34 \text{ cm} + (A) \ 29 \text{ cm} = 63 \text{ cm}$$

The most frequently used ratio on the other hand is:

$$2 \times (S) \ 18 \text{ cm} = 36 \text{ cm} + (A) \ 27 \text{ cm} = 63 \text{ cm}$$

An acceptable rise/run ratio for domestic stairs is:

$$2 \times (S) \ 19 \text{ cm} = 38 \text{ cm} + (A) \ 25 \text{ cm} = 63 \text{ cm}$$

For secondary stairs (see also DIN 18065), the following ratio is still permissible:

$$2 \times (S) \ 21 \text{ cm} = 42 \text{ cm} + (A) \ 21 \text{ cm} = 63 \text{ cm}$$

In our subsequent calculations an attempt will be made to achieve the most frequently used ratio of:

$$2 \times (S) \ 18 \text{ cm} = 36 \text{ cm} + (A) \ 27 \text{ cm} = 63 \text{ cm}$$

3.1.2.1. Design of rise/run dimensions on the basis of given floor heights and basic dimensions

When designing a flight of stairs, one tread should always be deducted from the number of rises; i.e., with five rises the calculations should be based on four treads. Before beginning the calculations, the number of rises must be determined, with 18.00 cm being the rise height.

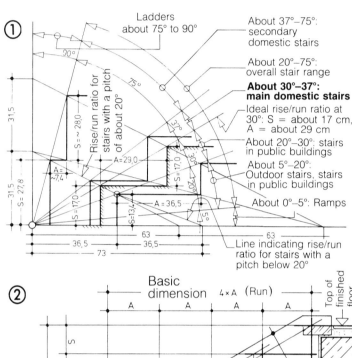

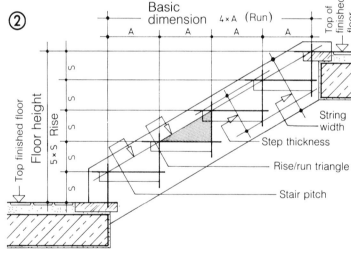

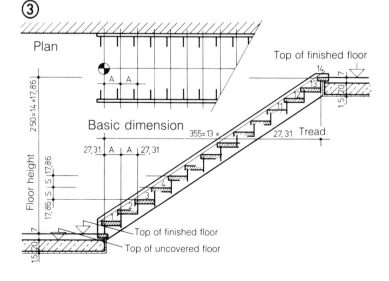

Example 1:
Floor height: 2.50 m. Basic dimension: 3.55 m.
1. Assumption:

Total rise: 250.00 cm ÷ 18 = (13.88) 14 rises
No fractions of rises are permitted, and results are rounded up or down. Our example is therefore based on 14 rises. On this basis the exact height of the rises is calculated.
Total rise: 250.00 ÷ 14 = 17.86 cm per rise (S)
With 14 rises, the calculation of the tread width is based on 13 treads:
Tread depth: 355.00 ÷ 13 = 27.30 cm per tread (A)

Check of rise/run ratio:

Ratio: 2 × (S) 17.86 = 35.72 + (A) 27.31 = 63.03 cm.
The rise/run ratio is ideal, and the staircase can be designed and built accordingly.

Example 2:
Floor height: 2.72 m. Basic dimension: 2.94 m.
1. Assumption:

Total rise: 272.00 ÷ 18 = (15.11) 15 rises
Height of rises: 272.00 ÷ 15 = 18.13 cm per rise
Tread depth 294.00 ÷ 14 = 21.00 cm per tread
Rise/run ratio: 2 × 18.13 = 36.26 + 21.00 = 57.26 cm.

The rise/run ratio is too small. In order to come closer to the ideal figure of 63, the number of rises should be reduced by one.

2. Assumption:

Height of rises: 272.00 ÷ 14 = 19.43 cm per rise
Tread depth: 294.00 ÷ 13 = 22.62 cm per tread
Rise/run ratio: 2 × 19.43 = 38.86 + 22.62 = 61.48 cm.

In order to make sure that it is not possible to find a more advantageous rise/run ratio for this by now steep staircase, another set of calculations should be made in which the number of rises is again reduced by one.
3. Assumption:

Height of rises: 272.00 ÷ 13 = 20.92 cm per rise
Tread depth: 294.00 ÷ 12 = 24.50 per tread
Rise/run ratio: 2 × 20.92 = 41.84 + 24.50 = 66.34 cm.

The rise/run ratio of 61.48 cm of the second set of calculations is closer to the ideal figure of 63.00 than that of the third assumption, which results in 66.34 cm.

Example 3:
Floor height: 0.78 m. Basic dimension: 0.82 m.
1. Assumption:

Total rise: 78.00 cm ÷ 18 = (4.33) 4 rises
Height of rises: 78.00 ÷ 4 = 19.50 cm per rise
Tread depth: 82.00 ÷ 3 = 27.33 cm per tread
Rise/run ratio: 2 × 19.50 = 39.00 + 27.33 = 66.33 cm.

This ratio is too high. It will be reduced if, in Assumption 2, the number of rises is reduced by one.
2. Assumption:

Height of rises: 78.00 ÷ 5 = 15.60 cm per rise
Tread depth: 82.00 ÷ 4 = 20.50 cm per tread
Rise/run ratio: 2 × 15.50 = 31.00 + 20.50 = 51.50 cm.

This ratio is far below 63 cm so that the result of Assumption 1 is more favorable, as long as it is not possible to increase the tread depth.

3.1.2.2.   Calculation of the basic stair dimensions on the basis of the given floor height
The floor height must be known whenever stairs are to be designed. The height is divided by the number of rises, with the standard height being taken as 18 cm. On this basis, the exact height is calculated. One then deducts twice the calculated rise height from 63 cm, the ideal rise/run ratio, so that one obtains the correct tread depth.
This is then multiplied by the number of rises; it must be borne in mind that one rise fewer is used in these calculations (for example: with 15 rises there are 14 tread depths).

Example 1:

Floor height: 2.50 cm. Basic dimension: ?
No. of rises: 250.00 cm ÷ 18 = (13.88) 14 rises
Height per rise: 250.00 cm ÷ 14 = 17.86 cm per rise
Tread depth: 63.00 − (2 × 17.86) = 35.72
    = (27.28) 27 cm per tread
Basic dimension: (14 − 1 =) 13 × 27.00 cm = 351.00 m
Rise/run ratio: 2 × 17.86 = 35.72 + 27.00 = 62.72

(Check)
The length of the basic dimension is 3.50 m.

Example 2:

Floor height: 2.82 m. Basic dimension: ?
No. of rises: 282.00 cm ÷ 18 = (15.66) 16 rises
Height per rise: 282.00 cm ÷ 16 = 17.63 cm per rise
Tread depth: 63.00 − (2 × 17.63) = 35.26 = (27.74) 28 cm
Basic dimension: (16 − 1) 15 × 28.00 = 4.10 cm
Rise/run ratio: 2 × 17.63 = 35.26 + 28.00 = 63.26

(Check)
The basic dimension is 4.20 m.

Example 3:

Floor height: 0.95 m. Basic dimensions: ?
No. of rises: 98.00 ÷ 18 = (5.44) 5 rises
Height per rise: 98.00 cm ÷ 5 = 19.60 cm per rise
Since 19.60 cm is too high, new calculations based on 6 rises have to be carried out.
Height of rise: 98.00 cm ÷ 6 = 16.33 cm per rise
Tread depth: 63 cm − (2 × 16.33) = 32.66 = (30.24) 30.00 cm
Basic dimension: (6 − 1) = 5 × 30.00 = 1.50 m
Rise/run ratio: 2 × 16.33 = 32.66 + 30.00 = 62.66 cm

(Check)
The basic dimension is 1.50 m.

3.1.2.3.   Calculation of rise/run ratio with a stair pitch of less than 20°

Measured on plan, the length of the average pace is 70–75 cm, the average length being 73 cm (1). The steeper the surface the shorter the step, which is reduced to about 63 cm at an angle of about 20°. If the angle is less than 20° it is not possible to make a formula-based calculation. As a result, a number of rise/run ratios were determined graphically in fig. 4. Intermediate values can be obtained by interpolation.

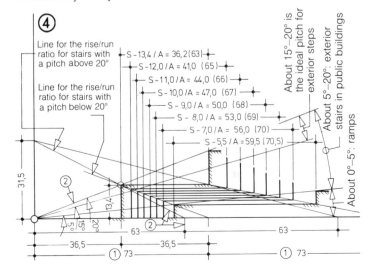

3.1.2.4. Rise/run tables

Rise/run tables are a time-saving aid when designing stairs; they can be used in the following cases:

1. Determination of the rise and tread dimensions in the case of given floor and basic dimensions.
2. The determination of rise and tread dimensions and the determination of basic dimensions with known floor heights.

Example for case No. 1:

Given floor height: 2.50 m.
Given basic dimension: 3.68 m.
Wanted: Number of rises and treads as well as height of rise and depth of tread.

One looks for the figure range that comprises the basic dimension 368 among the three sets of horizontal figures belonging to the floor height of 250. This is found in the second line between 324 and 384 (basic dimension). The same line gives the number of rises as 14, the height of rises as 17.86 cm, and the number of treads as 13. If one divides the given figure for the basic dimension by 13 (368 ÷ 13 = 28.31), one finds the depth of tread: 28.31 cm.

The stair data in example 1 are thus:

Floor height: 250 cm = 14 × 17.86 cm rise
Basic dimension: 368 cm = 13 × 28.31 cm tread
Rise/run ratio: 2 × 17.86 = 35.72 + 28.31 = 63.03 cm.

Example for case No. 2:

Given floor height: 2.75 m.
Wanted: Number of rises and treads, height of rises and depth of treads, as well as basic dimension of the staircase.

We look for the given dimension of 2.75 m in the column showing floor heights. Three sets of stairs of different pitches are dimensioned in three horizontal lines. The dimensions of the middle line are those for a standard staircase, the dimensions of the first line the most comfortable, and the dimensions of the third line a still acceptable domestic staircase. We choose the dimensions for the most comfortable domestic staircase:

Floor height: 275 = 16 × 17.19 cm per rise
Basic dimension: 429 = 15 × 28.62 cm per tread
Rise/run ratio: 2 × 17.19 = 34.38 + 28.62 = 63

Floor height	No. of rises	Height of rises	No. of treads	Depth of treads	Basic dimension	Basic dimension range
250	13	19.23	12	24.54	294	270-323
	14	17.86	13	27.29	355	324-383
	15	16.67	14	29.67	415	384-443
251	13	19.31	12	24.38	293	269-321
	14	17.93	13	27.14	353	322-382
	15	16.73	14	29.53	413	383-441
252	13	19.38	12	24.23	291	267-319
	14	18.00	13	27.00	351	320-380
	15	16.80	14	29.40	412	381-440
253	13	19.46	12	24.08	289	265-317
	14	18.07	13	26.86	349	318-378
	15	16.87	14	29.27	410	379-438
254	13	19.54	12	23.92	287	263-315
	14	18.14	13	26.71	347	316-376
	15	16.93	14	29.13	408	377-436
255	13	19.62	12	23.77	285	261-314
	14	18.21	13	26.57	345	315-374
	15	17.00	14	29.00	406	375-434
256	13	19.69	12	23.62	283	259-312
	14	18.29	13	26.43	344	313-372
	15	17.07	14	28.87	404	373-432
257	13	19.77	12	23.46	282	258-310
	14	18.36	13	26.29	342	311-370
	15	17.13	14	28.73	402	371-430
258	13	19.85	12	23.31	280	256-308
	14	18.43	13	26.14	340	309-369
	15	17.20	14	28.60	400	370-428
259	13	19.92	12	23.15	278	254-306
	14	18.50	13	26.00	338	307-367
	15	17.27	14	28.47	399	368-427
260	13	20.00	12	23.00	276	252-304
	14	18.57	13	25.86	336	305-365
	15	17.33	14	28.33	397	366-425

Floor height	No. of rises	Height of rises	No. of treads	Depth of treads	Basic dimension	Basic dimension range
261	14	18.64	13	25.71	334	308-363
	15	17.40	14	28.20	395	364-424
	16	16.31	15	30.38	456	425-486
262	14	18.71	13	25.57	332	306-361
	15	17.47	14	28.07	393	362-422
	16	16.38	15	30.25	454	423-484
263	14	18.79	13	25.43	331	305-359
	15	17.53	14	27.93	391	360-420
	16	16.44	15	30.13	452	421-482
264	14	18.86	13	25.29	329	303-357
	15	17.60	14	27.80	389	358-418
	16	16.50	15	30.00	450	419-480
265	14	18.93	13	25.14	327	301-355
	15	17.67	14	27.67	387	356-416
	16	16.56	15	29.88	448	417-478
266	14	19.00	13	25.00	325	299-354
	15	17.73	14	27.53	385	355-414
	16	16.63	15	29.75	446	415-476
267	14	19.07	13	24.86	323	297-352
	15	17.80	14	27.40	384	353-412
	16	16.69	15	29.63	444	413-474
268	14	19.14	13	24.71	321	295-350
	15	17.87	14	27.27	382	351-411
	16	16.75	15	29.50	443	412-473
269	14	19.21	13	24.57	319	293-348
	15	17.93	14	27.13	380	349-409
	16	16.81	15	29.38	441	410-471
270	14	19.29	13	24.43	318	292-346
	15	18.00	14	27.00	378	347-407
	16	16.88	15	29.25	439	408-469
271	14	19.36	13	24.29	316	290-344
	15	18.07	14	26.87	376	345-405
	16	16.94	15	29.13	437	406-467

Floor height	No. of rises	Height of rises	No. of treads	Depth of treads	Basic dimension	Basic dimension range
272	14	19.43	13	24.14	314	288-342
	15	18.13	14	26.73	374	343-403
	16	17.00	15	29.00	435	404-465
273	14	19.50	13	24.00	J12	286-341
	15	18.20	14	26.60	372	342-401
	16	17.06	15	28.88	433	402-463
274	14	19.57	13	23.86	310	284-339
	15	18.27	14	26.47	371	340-399
	16	17.13	15	28.75	431	400-461
275	14	19.64	13	23.71	308	282-337
	15	18.33	14	26.33	369	338-397
	16	17.19	15	28.63	429	398-459
276	14	19.71	13	23.57	306	280-335
	15	18.40	14	26.20	367	336-396
	16	17.25	15	28.50	428	397-458
277	14	19.79	13	23.43	305	279-333
	15	18.47	14	26.07	365	334-394
	16	17.31	15	28.38	426	395-456
278	14	19.86	13	23.29	303	277-331
	15	18.53	14	25.93	363	332-392
	16	17.38	15	28.25	424	393-454
279	15	18.60	14	25.80	361	333-390
	16	17.44	15	28.13	422	391-451
	17	16.41	16	30.18	483	452-515
280	15	18.67	14	25.67	359	331-388
	16	17.50	15	28.00	420	389-449
	17	16.47	16	30.06	481	450-513
281	15	18.73	14	25.53	357	329-386
	16	17.56	15	27.88	418	387-447
	17	16.53	16	29.94	479	448-511
282	15	18.80	14	25.40	356	328-384
	16	17.63	15	27.75	416	385-445
	17	16.59	16	29.82	477	446-509
283	15	18.87	14	25.27	354	326-383
	16	17.69	15	27.63	414	384-443
	17	16.65	16	29.71	475	444-507
284	15	18.93	14	25.13	352	324-381
	16	17.75	15	27.50	413	382-441
	17	16.71	16	29.59	473	442-505
285	15	19.00	14	25.00	350	322-379
	16	17.81	15	27.38	411	380-440
	17	16.76	16	29.47	472	441-504
286	15	19.07	14	24.87	348	320-377
	16	17.88	15	27.25	409	378-438
	17	16.82	16	29.35	470	439-502
287	15	19.13	14	24.73	346	318-375
	16	17.94	15	27.13	407	376-436
	17	16.88	16	29.24	468	437-500
288	15	19.20	14	24.60	344	316-373
	16	18.00	15	27.00	405	374-434
	17	16.94	16	29.12	466	435-498
289	15	19.27	14	24.47	343	315-371
	16	18.06	15	26.88	403	372-432
	17	17.00	16	29.00	464	433-496
290	15	19.33	14	24.33	341	313-369
	16	18.13	15	26.75	401	370-430
	17	17.06	16	28.88	462	431-494
291	15	19.40	14	24.20	339	311-368
	16	18.19	15	26.63	399	369-428
	17	17.12	16	28.76	460	429-492
292	15	19.47	14	24.07	337	309-366
	16	18.25	15	26.50	398	367-426
	17	17.18	16	28.65	458	427-490
293	15	19.53	14	23.93	335	307-364
	16	18.31	15	26.38	396	365-425
	17	17.24	16	28.53	456	426-488
294	15	19.60	14	23.80	333	305-362
	16	18.38	15	26.25	394	363-423
	17	17.29	16	28.41	455	424-487
295	15	19.67	14	23.67	331	303-360
	16	18.44	15	26.13	392	361-421
	17	17.35	16	28.29	453	422-485
296	15	19.73	14	23.53	329	301-358
	16	18.50	15	26.00	390	359-419
	17	17.41	16	28.18	451	420-483
297	16	18.56	15	25.88	388	358-417
	17	17.47	16	28.06	449	418-478
	18	16.50	17	30.00	510	479-544
298	16	18.63	15	25.75	386	356-415
	17	17.53	16	27.94	447	416-476
	18	16.56	17	29.89	508	477-542
299	16	18.69	15	25.63	384	354-413
	17	17.59	16	27.82	445	414-474
	18	16.61	17	29.78	506	475-540
300	16	18.75	15	25.50	383	353-411
	17	17.65	16	27.71	443	412-472
	18	16.67	17	29.67	504	473-538
301	16	18.81	15	25.38	381	351-410
	17	17.71	16	27.59	441	411-471
	18	16.72	17	29.56	502	472-536
302	16	18.88	15	25.25	379	349-408
	17	17.76	16	27.47	440	409-469
	18	16.78	17	29.44	501	470-535
303	16	18.94	15	25.13	377	347-406
	17	17.82	16	27.35	438	407-467
	18	16.83	17	29.33	499	468-533
304	16	19.00	15	25.00	375	345-404
	17	17.88	16	27.24	436	405-465
	18	16.89	17	29.22	497	466-531
305	16	19.06	15	24.88	373	343-402
	17	17.94	16	27.12	434	403-463
	18	16.94	17	29.11	495	464-529
306	16	19.13	15	24.75	371	341-400
	17	18.00	16	27.00	432	401-461
	18	17.00	17	29.00	493	462-527
307	16	19.19	15	24.63	369	339-398
	17	18.06	16	26.88	430	399-459
	18	17.06	17	28.89	491	460-525

3.2. WALKING LINE

3.2.1. Function of the walking line

3.2.2.1. Calculation of the rise/run ratio for spiral staircases with two-way traffic

1. Calculation of the rise/run ratio at the centrally arranged (auxiliary) walking line (L.m.):

Floor height: 1.78 m $= 11 \times 16.18$ cm rise
Basic dimension: 3.11 m $= 10 \times 31.10$ cm tread
Rise/run ratio: $2 \times 16.18 = 32.26 + 31.10 = 63.46$ cm.

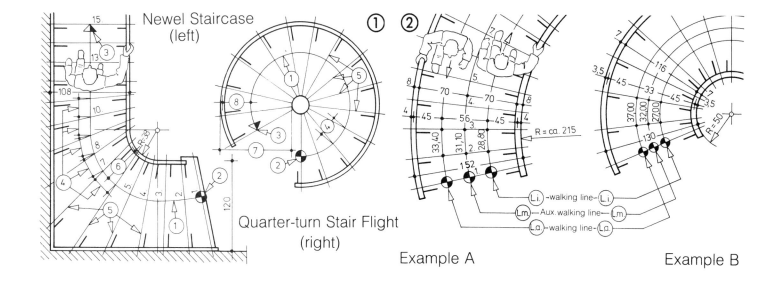

Newel Staircase (left) Quarter-turn Stair Flight (right)

① ②

L.i. -walking line- L.i.
Lm -Aux. walking line- Lm
L.a. -walking line- L.a.

Example A   Example B

The walking line (line 1) is an imaginary line indicating the path usually taken by stair users. It is indicated in all staircase plans. Its limits are marked at the bottom of the stairs by a circle (2), at the top of the stairs by an arrow (3). It thus serves to indicate both the direction of inclination of the stairs and whether they are left or right stairs. The length of the walking line thus indicated is at the same time the basic dimension on which the rise and tread design is based. The calculated tread depths (4) are marked along the walking line, and the shape of the steps is determined on the basis of these intersections (5) (see also section on tapering of treads).

The distance between the center of the handrail and the walking line (6) is about 45 cm with stairs of standard width. With stair widths below 90 cm, the walking line should be in the center of the steps.

In the case of newel stairs with a width (7) of more than 75 cm, a distance (8) of about 37 cm between the center of the handrail and the walking line should be marked; with even narrower staircases the walking line should be indicated in the center (between the center of the handrail and the edge of the newel post).

3.2.2. Two walking lines on stairs with two-way traffic

In the case of wider staircases subject to two-way traffic and with a handrail on either side, one finds, in practice, two walking lines that are parallel to the handrails. In examples A and B of fig. 2 these are indicated as L.i. (inside walking line) and L.a. (outside walking line). When designing such a staircase, the rise/run ratio along the auxiliary (imaginary) walking line in the center of the staircase (L.m.) should be as close as possible to the ideal figure of 63 cm. Afterwards, the ratio of the actual walking lines should be calculated, and it must be borne in mind that the ratio should not be less than 60 cm for the inner line or more than 66 cm for the outer line.

2. Calculation of the inner walking line (L.i.):

Floor height: 1.80 m $= 11 \times 16.18$ cm rise
Basic dimension: 2.88 m $= 10 \times 28.8$ cm tread
Rise/run ratio: $2 \times 16.18 = 32.36 + 28.8 = 60.16$ cm.

3. Calculation of the outer walking line (L.a.):

Floor height: 1.80 m $= 11 \times 16.18$ cm rise
Basic dimension: 3.34 m $= 10 \times 33.40$ cm tread
Rise/run ratio: $2 \times 16.18 = 32.36 + 33.40 = 65.76$ cm.

These calculations have shown that the inner diameter of such a curved staircase for two-way traffic must not be too small.

In example B of fig. 3, this mistake has been made. The central walking line has a ratio of ($2 \times 15.50 + 32 =$) 63 cm.

The inner walking line has a rise/run ratio of ($2 \times 15.50 = 31.00 + 27.00 =$) 58.00 and the outer walking line one of ($2 \times 15.50 = 31.00 + 37.00 =$) 68.00. According to our rule, all ratios should be between 60 and 66 cm. Such a staircase will therefore not be comfortable to negotiate.

If, on the other hand, compelling reasons exist for the construction of a staircase where our rules cannot be adhered to, care should be taken that on the inner line (L.i.), at least, the ratio is no less than 58 cm so that no hazard exists along this line.

When designing a staircase intended for two-way traffic, the inner radius should be graphically determined, bearing in mind the correct rise/run ratio.

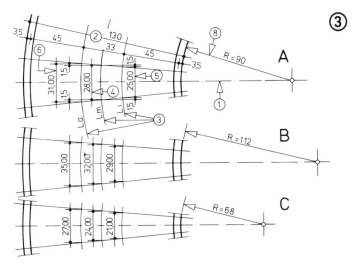

③

Walking Line

Walking Area

ca. 45 cm.

ca. 45 cm.

Example:

Starting with a baseline A (1), the stair width (2) and the walking lines (3) are entered at right angles to (1). The calculated tread depth of 28 cm in the center of the stairs (L.m.) (rise/run ratio $2 \times 17.50 = 35.00 + 28.00 = 63.00$) is marked so that it intersects and halves the base line (4). The minimum depth at L.i. is now found to be ($28.00 - 3$ cm =) 25 cm (5), since ($2 \times 17.50 = 35.00 + 25 =$) 60 cm, this being the lower limit of the known rise/run ratio. The upper limit is about 66 cm. As a result, ($28.00 + 3.00 =$) 31 cm is now transferred to L.a. (6). The intersections (7) are marked on the base line, so that we obtain an inner radius for the staircase of, in our example, 90 cm (8). Example 3 shows, under B, that with a mean tread depth of 32 cm, the inner radius staircase is 1.12 m; in the case of C and a tread of 24 cm, the radius is 68 cm.

Figure 4 shows the plan of a representative semicircular staircase. Rise/run ratio at the center:

$$2 \times 15.2 = 30.4 + 32.6 = 63.0 \text{ cm.}$$

By providing landings, a rise/run ratio of ($2 \times 15.2 = 30.4 + 36.5 =$) 66.9 can still be obtained. The landings are dimensioned so that one normal pace can be taken (see landings). The rise/run ratio of the inner walking line (L.i.) is ($2 \times 15.2 = 30.4 + 28.8 =$) 59.2 cm and thus acceptable.

In the square staircase shown in fig. 5, example D, landings have been provided in the corners. As a result, an acceptable rise/run ratio is obtained along the outer walking line (L.a.) and the inner and outer walking lines are comfortable.

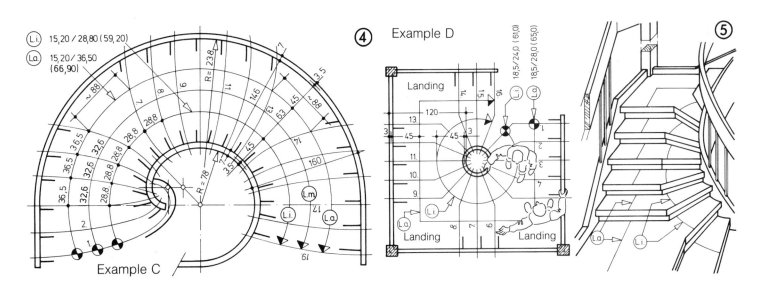

④ Example C

⑤ Example D

27

3.3. FLIGHT WIDTHS OF A STAIRCASE

The flight width of a staircase equals the design width of a staircase (A-1), while the utilizable flight width equals the clear width of the built staircase measured at the level of the handrail (A-2).

The most important flight widths for domestic buildings are laid down in DIN 18065 as follows:

Type of Staircase	Utilizable width
Stairs in single-family units	800 mm
Basement stairs, attic stairs, secondary stairs (service stairs), emergency stairs, stairs in multifamily units with 2 or 3 dwellings	900 mm
Stairs in multifamily units with more than 3 dwellings	1,000 mm
Stairs in domestic high-rise buildings	1,250 mm

Secondary space-saving stairs must have a clear width of at least 60 cm. Where it is necessary to build stairs sufficiently wide for two persons to pass each other, the clear width should be 1.30 m but no less than 1.10 m; for three persons, the corresponding figures are 1.90 m and 1.70 m.

In the case of stairs that are circular in the lower section, it is permissible, for design reasons, for the lower part to be wider than the remainder of the staircase; in this case the walking line must run parallel to the handrail (D + E). It is also permissible to vary the flight width in the area of a landing by 10 to 20 cm (fig. C).

3.4. HEADROOM

According to DIN 18065, Staircases in Domestic Buildings, the headroom should be 2.10 m (1).

When negotiating stairs with a pitch of more than 45°, where it may be assumed that the body is inclined at an angle of about 10°, the headroom should still be at least 2.00 m (2). In the case of wider staircases, it is desirable, for optical reasons, to add the amount by which the stair width exceeds 1.00 m to the headroom of 2.10 m.

Example:

Flight width: 125 cm
Headroom: 210 cm + 25 = 235 cm.

Wall lights should not interfere with stair traffic.

In the case of flights of stairs arranged above one another, the headroom is calculated on the basis of the stair depth measured both vertically and at right angles to the pitch of the stair flight.

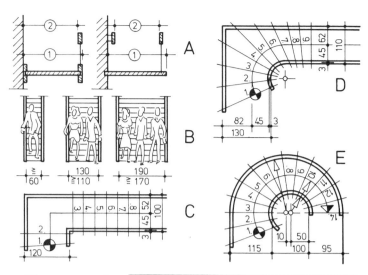

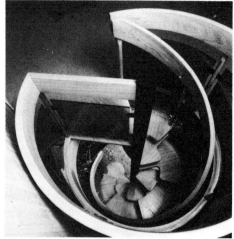

The newel staircase is only about 60 to 65 cm wide between newel and handrail. The walking line used by the feet thus becomes fixed, ensuring the safe negotiation of this steep and most space-saving of staircases.

3.5. FLOOR OPENINGS FOR STAIRCASES

3.5.1. Determination of the amount of ceiling stagger

The amount of stagger of flights of stairs arranged above one another can be calculated on the basis of the predetermined headroom. The illustration below provides a graphic solution; the upper and lower edges of the steps are connected by lines indicating the stair pitch (1), the headroom dimension of 2.10 m is marked vertically above and below these lines (2), and the line indicating the next flight is drawn parallel to that marking the edges of the flight of stairs above and below, respectively (3). The intersection (4) with the underside of the ceiling indicates the edge of the ceiling and the amount by which the ceilings are staggered (5). The intersection (6) of the line indicating the staircase edge of the lower flight of stairs and the finished floor of the ground floor (E) marks the floor opening at ground floor level; the dimension (7) must also be taken into account.

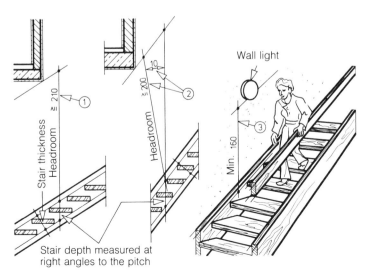

Stair depth measured at right angles to the pitch

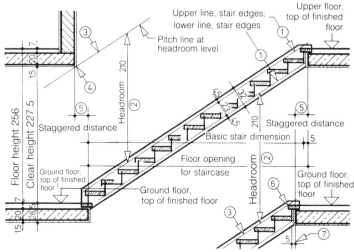

3.6. PLANS AND ELEVATIONS OF SPIRAL STAIRS

3.6.1. Stair plans

Circular staircases obtain their momentum from the plan and shape of the properly or incorrectly "tapered" stair treads.

In example (A) too many straight steps have been provided. As a result, the central section of the string shown in the string ramp (A3) is too steep in the area of the string wreath (1), while the adjacent strings (2) are too shallow. The discontinuation of the flowing "movement" arises because the steps fitted into the strings are all of equal depth but become narrower in the area of the string wreaths.

The smallest width arising at the corner steps 7 and 10 along the wall side (A2) should not be less than 10 cm (4) so that acute angles, in which dirt can accumulate, are avoided.

Example (B) shows a similar plan where these deficiencies have been avoided.

In example (B 1) the proportion of tapered steps is much greater than in (A1). In the area where step flight is straight, it should amount approximately to the width of the steps, so that the tapered section amounts to roughly twice the stair width, starting from the end wall (5). The tapered section of the staircase thus extends from step 2 to step 15.

In order to achieve a smooth flow on the string ramp (B3), care must be taken that the depth of the steps along the face string increases and decreases evenly. (Various tapering methods will be dealt with at a later stage.)

If the corner steps 6 and 10 are to have approximately the same dimensions on the wall side (6) and (7), a wedge-shaped step (step 8 in fig. B2) should be provided. Finally, it is suggested that the solid rectangular step (step 1) should be slightly tapered toward the inside. This makes it more inviting.

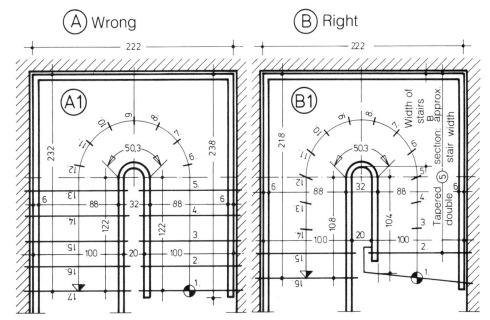

Ⓐ Wrong   Ⓑ Right

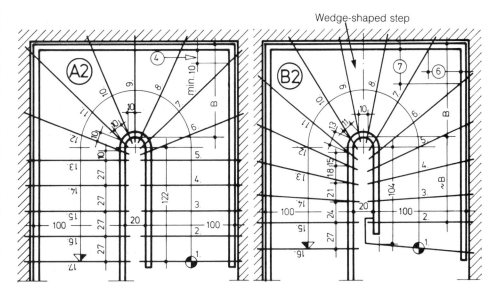

Wedge-shaped step

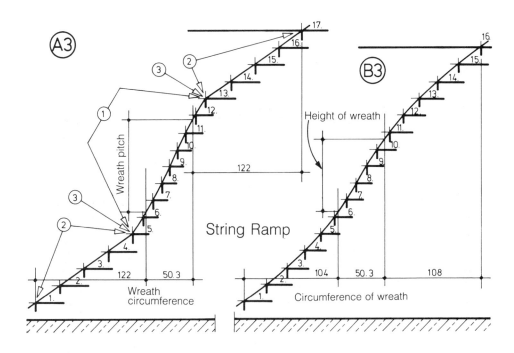

Ⓐ3   Ⓑ3   String Ramp

29

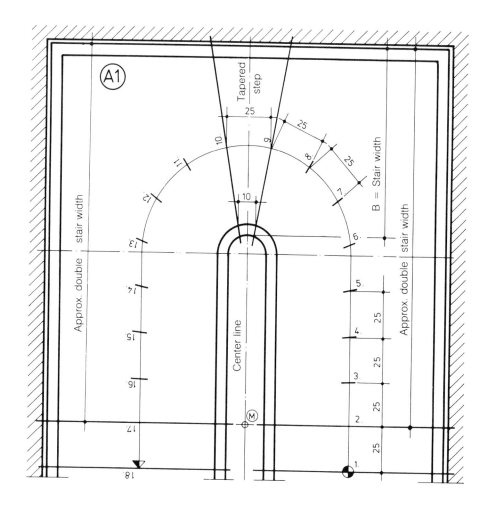

3.6.2. Tapering of steps in circular staircases

All tapering methods must lead to the following result: the step depth along the face string must increase and decrease evenly in order to produce a smoothly flowing facing. Whichever tapering method is chosen, once the plan has been decided upon and the tread depth determined, one determines the start of the circular section. As an aid to this process one finds out which step lies twice the stair width away from the facing wall. In example (A1) this distance coincides with the front edges of steps 2 and 17. At this early stage, when the arrangement of steps along the walking line is decided upon, it is advisable to provide for a wedge-shaped step, which ensures advantageous positioning of corner steps.

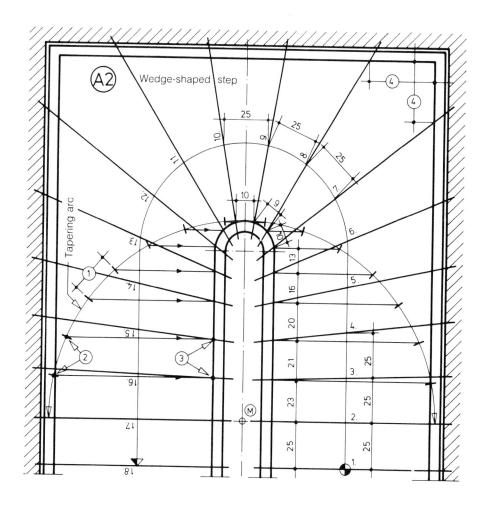

3.6.2.1. Tapering of steps using the circular arc method

In figure (A2), steps 2 and 17 mark the beginning and end of the tapered section. These lines intersect the center line in the stairwell where they form the center point (M) for the tapering arc. The line of this semicircle is divided into fifteen equal sections (1); i.e., the number of steps and the fifteen points thus gained are connected horizontally with the line indicating the face string (2). These intersections (3) are connected with the points on the walking line. This completes the tapering process. If one compares the depth of steps 1 and 9, one finds that the result is satisfactory. The corner steps are advantageously placed in the wall-string corners (4).

In example (B), the circular arc method has been used for a flight of stairs with a quarter turn in the lower section.

Here too, the tapered section between steps 1 and 8 has already been designed (fig. B1). When tapering was carried out (B2), step 3 was almost exactly in the corner of the wall string. This was avoided in (C1) when the front edge of the first step was moved by about 15 cm (1). Since, in the new example, step 1 no longer forms part of the tapering arc, its depth (14 cm) should be determined first. The dimension of 14 cm was found empirically in example (B2) (2). The center point of the tapering arc (M) marks the intersection of the extended edges of steps 2 and 9. The method of tapering has already been described. The corner step is now well placed across the corner of the wall string.

A dimensional check of tread depth along the face string shows that the difference between steps is between 2 and 3 cms. This even increase ensures an ideal flow of the face string.

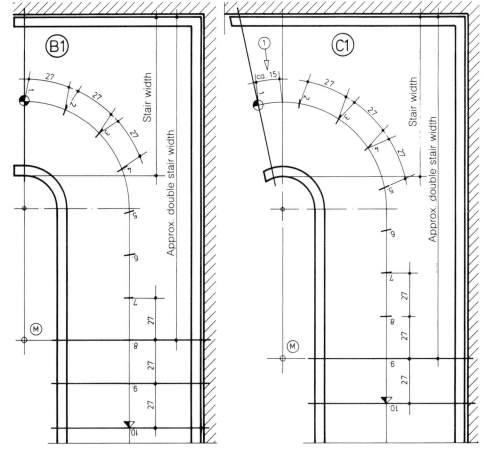

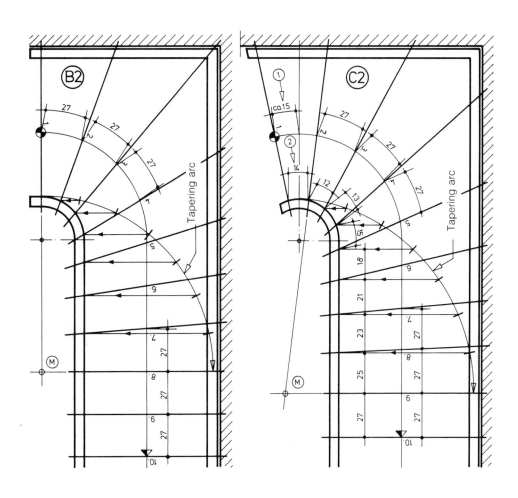

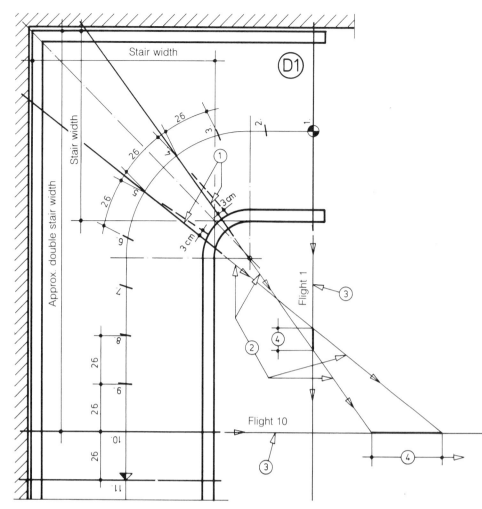

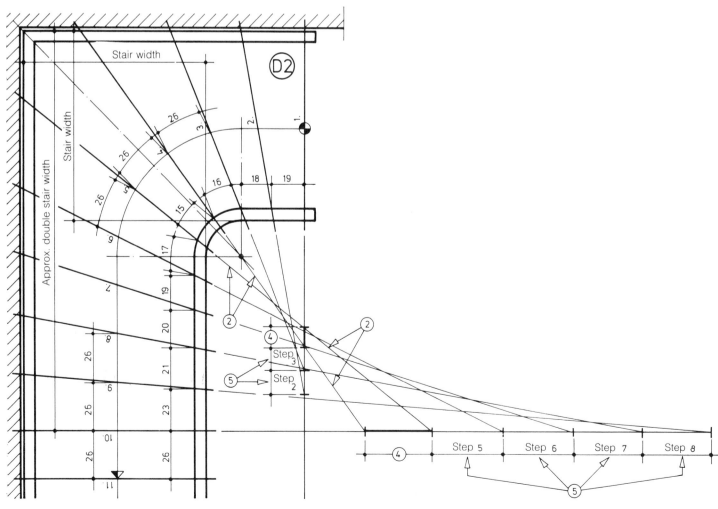

3.6.2.2. Tapering of steps using the flight line method

With this method one again determines, as a first step, the tapered section in the upper part of the staircase (D1). Here, step 10 is closest to the double stair width. At the bottom of the stairs, step 1, which lies near a string wreath, has already been determined. In this case the corner step is moved back along the side of the face string by about 3 cm (1), to prevent steps 1 to 3 from becoming too narrow. The flight lines of the corner step 4 are extended in the direction of the face string (2) and beyond until they intersect the flight lines of steps 1 and 10 (3). The dimensions (4) obtained here are applied as many times on flight lines 1 and 10 as there are steps requiring tapering (D2). The resulting intersections are connected by lines with the indicated steps on the walking line. A dimensional check along the face string gives a satisfactory result.

This interesting flight-line method can also be applied to semicircular staircases as follows:

In the plan (fig. A) step 2 has been marked as the end of the circular section (1). Step 8 is marked as the wedge-shaped step (2). The tapering starts with the extension of the lines of step 8 (3) to the flight line of step 2 (4). The resulting dimension (5) is applied to the left as many times as steps need to be tapered. The intersections are connected with the points on the walking line.

The left-hand part of the staircase is tapered by the same method.

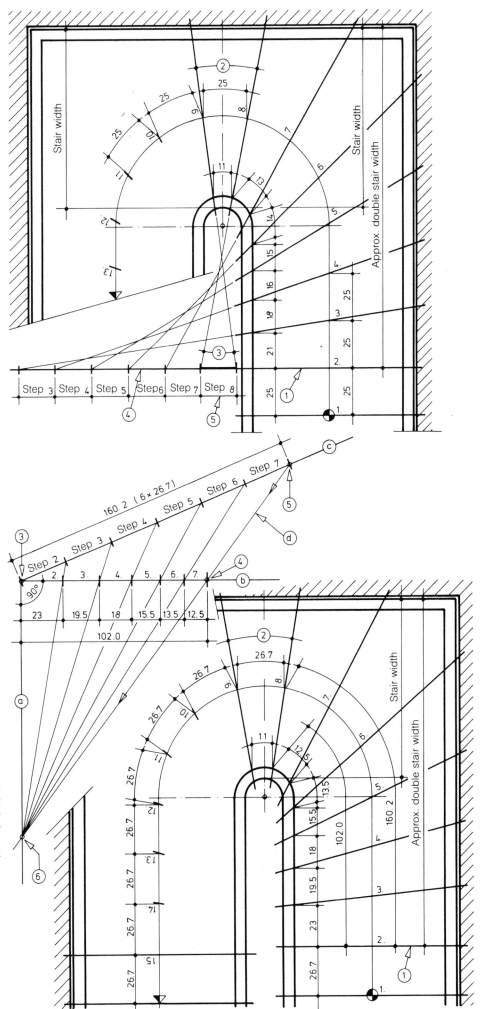
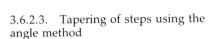

3.6.2.3. Tapering of steps using the angle method

As in the preceding example, here too step 2 (1) is the first to be in the tapered section. Again, step 8 is the wedge-shaped step (2). The total length of the steps requiring tapering is 160.2 cm measured on the walking line and 102.0 cm at the face string. On an additional elevation—suggested scale 1:10 —a right angle is drawn (a), (b) and above this an oblique line at an angle of approximately 20° (c). Starting at point (3) one enters along line (c) the walking line and stair dimensions. The length of the circular section of the face string of 102.0 cm should be marked on line (b). Line (d) is drawn through the intersections (4) and (5), which intersects line (a) at point (6). All the subdivisions for the individual steps on line (c) should now be connected with this point (6). The intersections on line (b) indicate the step depth, and these are to be used in the plan and indicated on the face-string side. The steps can subsequently be drawn in full.

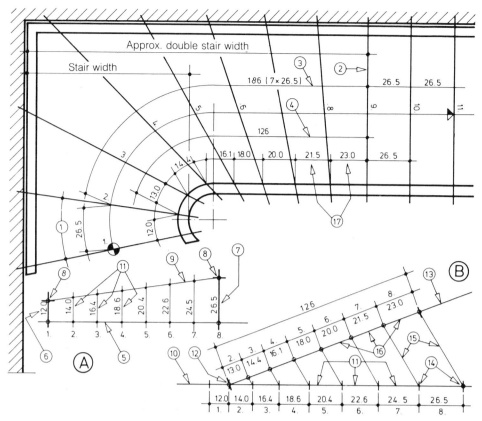

Here again, step 1 has been previously determined (1); subsequently, the tapered section has been limited to include step 9 (2). The walking line (3) measures 186 cm (7 × 26.5), the base line (4) at the face string 126 cm.

In a special elevation (A), as many equidistant vertical lines (about 15–20 cm) are drawn as steps that need tapering in addition to step 1. The basic dimensions of the first step are indicated on the first vertical line on the left (6) and the tread depth of 26.5 cm by the last step on the right-hand side (7). The resulting intersections (8) should be connected (9).

The widths (11) taken from fig. (A) are applied onto a new baseline, elevation (B) (10). On an additional line, starting from point (12) at an angle of about 20° (13), one marks the baseline length of the string of 126 cm and the intersections (14) by lines (15). One thus obtains the actual stair depths (16), and these can be entered in the plan. Subsequently, the tapered steps can be drawn (17).

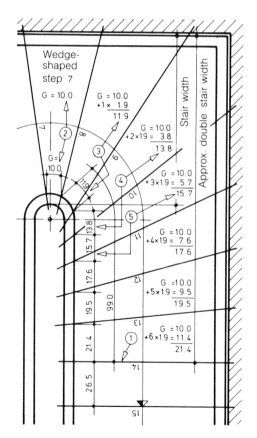

3.6.2.5. Tapering of steps using the mathematical method

When applying the mathematical method, the tapered section is again limited by the last straight step 14 (1) and the determination of the wedge-shaped step 7. The dimension of the wedge-shaped step 7 was provisionally assumed to be 10 cm (2), but this can be modified once the adjacent steps have been designed. This dimension is between 10 and 12 cm for semicircular stairs but becomes smaller with smaller string wreath diameters. (At this point DIN 18065 should be borne in mind: "In the case of circular staircases, a depth of tread of at least 100 mm must be provided at a distance of 150 mm from the narrowest point.") This basic dimension of 10 cm is now used for each step requiring tapering; i.e., for steps 8 to 13 = 6 × 10.0 = 60 cm. Consequently, 99.0 − 60.0 = 39.0 cm remains for fixed-length additions.

The numbers of additions are determined as follows:

Step 8	1 addition
Step 9	2 additions
Step 10	3 additions
Step 11	4 additions
Step 12	5 additions
Step 13	6 additions
Steps 8 to 13	21 additions

39.0 cm ÷ 21 additions = (1.86) 1.9 cm per addition.

This additional length of 1.9 cm should be added to the basic dimension in relevant multiples, for example:

For Step 8:

$$10 \text{ cm} + 1 \times 1.9 = 11.9 \text{ cm} \quad (3)$$

For Step 9:

$$10 \text{ cm} + (2 \times 1.9 =) 3.8 = 13.8 \text{ cm} \quad (4)$$

For Step 10:

$$10 \text{ cm} + (3 \times 1.9 =) 5.7 = 15.7 \text{ cm} \quad (5),$$

etc.

3.6.2.6. Tapering of steps using the freehand method

All tapering methods described so far lead to good results with string wreaths having openings over about 20 cm. Below that corrections are necessary, especially with steps that have very acute angles.

In this example the amount of tapering has been determined on the basis of the trapeze method. Although there is a regular increase in the depth of the steps on the face-string side (1), step 2 appears to be too narrow there (2). This observation leads to the following general rule: the steps of a tapered staircase show optimum tapering if the depth measured at right angles from the edge of the step increases or decreases evenly. In fig. (A) this has not been achieved. Although the length of the edge is adequate, it becomes apparent, when using the measurement method at right angles, that step 2, with 8 cm, is too narrow (2).

In fig. (B) the right-angle method has been applied, and the steps have been tapered accordingly. The result is better, although a check shows that the increase in depth along the face string is not even. This does not necessarily contradict the previously supported theory according to which the increase or decrease in depth along the face string should be uniform. For technical reasons it may be advantageous to place rods on the plan indicating the taper of the steps, as shown in fig. (C), and move them until the arrangement coincides with the principle of fig. (B). The width of these rods may, for example, equal the amount by which the tread projects over the riser, so that treads and risers can be determined at the same time. (In drawings with a scale of (or near) 1:10, commercially available fretsaw blades may prove useful.)

This "freehand" method of determining the taper is useful for a different reason, in that it permits the determination of the corner step on the wall side. In the case of string staircases, care should also be taken that, wherever possible, no risers or tread pitches should occur at the joint between string and string wreath.

Trapeze Method

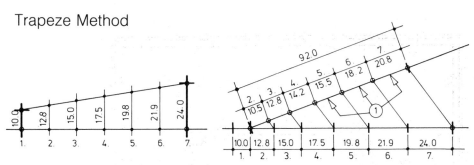

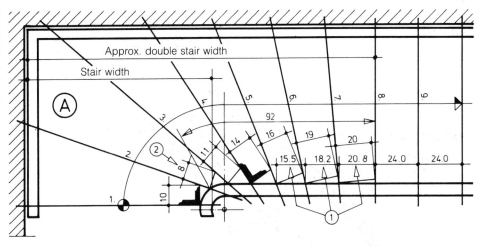

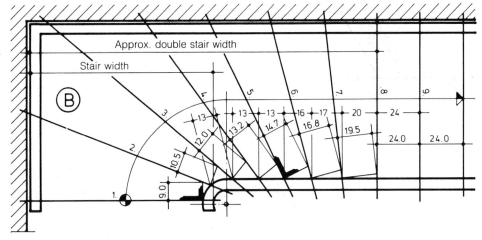

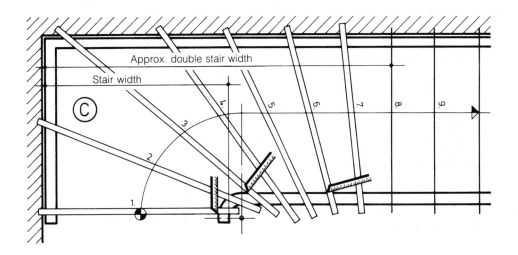

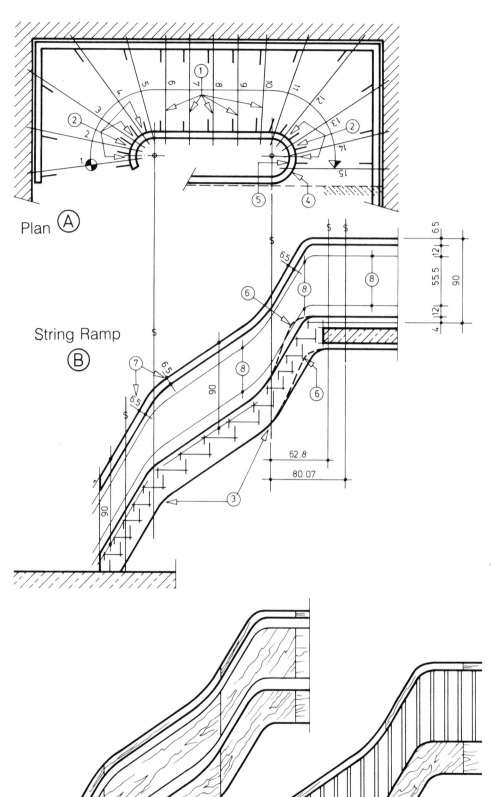

Plan Ⓐ

String Ramp Ⓑ

String Ramp Ⓒ
with Continuous Midrail

String Ramp Ⓓ
with Bannister Rods

3.6.3. Tapering of steps—Effects on the stair railing

The effects of good or bad tapering on the flow of the stair railing are demonstrated in the following examples. In plan (A) we find too many straight steps in the central section (1) of the flight of stairs. As a result, the steps are too narrow along the face string (2) in the circular section.

The line of the face string (B) is uneven and inelegant. A horizontal measurement shows that the arc of the string wreath at the top (4) measures 80.07 cm along the side of the stairs but only 62.80 cm on the outside (5). This means that on the face side the string-wreath section becomes even steeper. This is indicated in fig. (B) by the string ramp (6) and will be explained below with the help of a further example.

The intended handrail is molded and must be—measured at right angles (7)—of even thickness (6.5 cm). Measured from the front edge of the step, the railing is 90 cm high. The horizontal width of the midrail is calculated by deducting the permissible gap (2 × 12 cm), the string projection (4 cm), and the thickness of the handrail (6.5 cm) from the height of the railing (90 cm). It thus amounts to 55 cm. If one draws the midrail into the elevation it will be found that, although the vertical dimensions are approximately the same throughout (8), the width fluctuates greatly. Under these conditions it would be better not to have a solid midrail but to use bannister rods, as shown in fig. (D).

Plan (B) is a good example of well-tapered steps. The resulting string ramp thus shows an ideal line so that a midrail could easily be installed here.

A remarkable feature is that the width of step 14 (1) at the transitional string wreath on the face string exceeds that of step 13 (2). As a result, the arc leading from the inclined stair flight into the horizontal balustrade (3) becomes larger and more pleasing. If, instead of the molded handrailing (4), which must be parallel as shown in (5), a handrail with a rectangular cross section were to be installed, all dimensions could be marked with the help of a marking strip (6) as shown in fig. (B1). This would mean that the different widths (handrail, gaps, etc.) would become wider and narrower in the correct proportions.

In example (C), the tread widths along the face string are all identical (7) in order to achieve a straight pitch line for the face string (8) in the string ramp (C). The resulting advantages for the stair railing are so great that possible minor disadvantages in the plan can be ignored.

In order to obtain such a taper, step 1 has to be shifted along the face string from 125 cm (plan B) to 138 cm (plan C) (9). The types of stair railing indicated in the string ramp (C) are all very difficult to produce and should only be used for bannisters with a linear pitch line. In the case of saddle-type stairs with individually applied standards, their molding should be as shown in detail (C1).

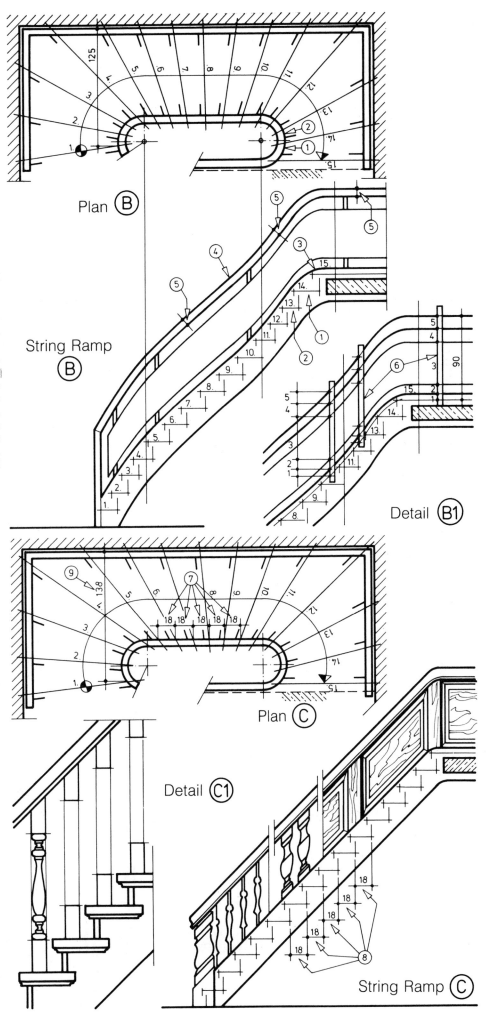

Plan (B)

String Ramp (B)

Detail (B1)

Plan (C)

Detail (C1)

String Ramp (C)

In all of the examples shown on this page, (D) to (J), the treads along the face string are identical. This can only be achieved with a string wreath whose radius is at least 30 cm and where minor dimensional changes can be made on the first and last steps.

Moreover, such tapering can only be done by the freehand method with the help of rods, by empirically shifting these and by taking frequent measurements. In the case of circular staircases with strings arranged as shown on page 39, the tapering of steps and the arrangement of standards go hand in hand.

In the example shown, the bottom step was deliberately extended by a considerable amount in order to give the stairs, for which only little space was available, the appearance of generous proportions.

Moreover, the relatively steep handrail above the face string was designed so that a smooth line was created, linking it visually with the horizontal helix on the same level as the standard on the wall side.

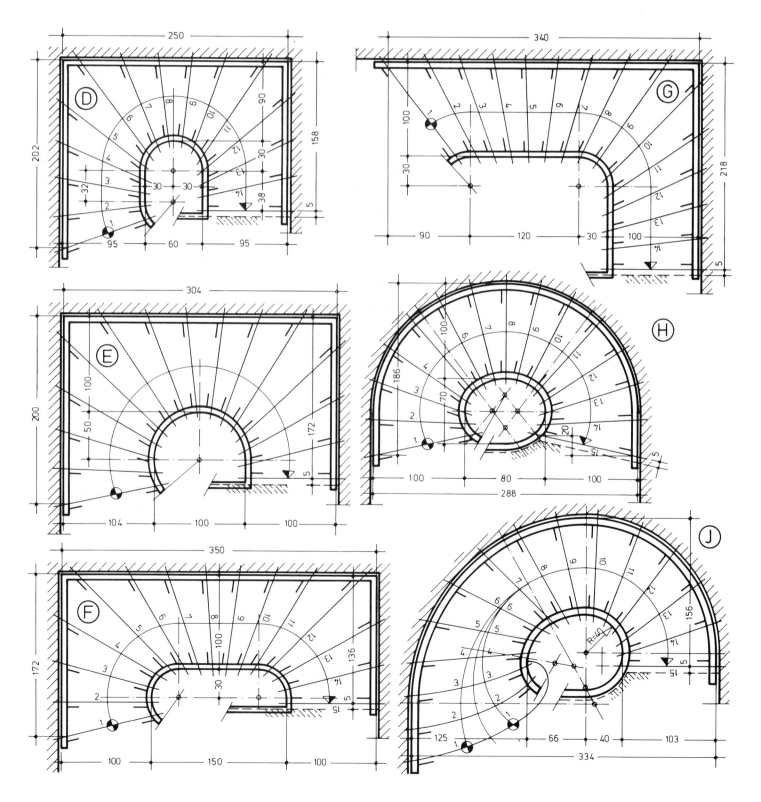

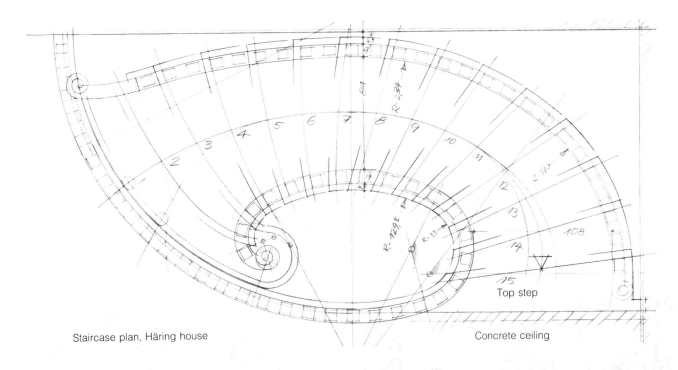

Staircase plan, Häring house

Top step

Concrete ceiling

Family coat of arms

Häring house,
handrail standard

Oberkochen, March 1976

3.7. STRING AND HANDRAIL WREATHS

3.7.1. Types of wreaths

There are basically two different types of wreaths:

1. Vertical wreaths
2. Horizontal wreaths

The vertical wreath (A), (1) normally extends from the lower edge of the string to the top of the handrail. The profile of the handrail is usually incorporated into it. This has the disadvantage that the grain of the handrail is interrupted (2). As a result, the top of the handrail in this section is formed by end grain. The grain of horizontal wreaths is parallel with that of strings and handrails (B). As a result, the direction of the grain within the railing is not interrupted.

3.7.2. Wreath designations

One distinguishes between wreaths connecting strings (B), (1) described as string wreaths, and those connecting handrail sections, known as handrail wreaths (2).

In fig. (C), a horizontal handrail wreath (1) has been fitted into a vertical string wreath (2), as shown by detail (3).

Depending on the shape of the plan, wreaths are described as quarter- or half-wreaths.

Legend:

(A) (1) Vertical quarter-wreath
(B) (1) Horizontal quarter-string wreath
     (2) Horizontal quarter-handrail wreath
(C) (1) Horizontal half-handrail wreath
     (2) Vertical half-string wreath
(D) (1) Horizontal half-handrail wreath
     (2) Horizontal half-string wreath
(E) (1) Horizontal half-string wreath— turning point
     (2) Horizontal half-handrail wreath— turning point
(F) (1) Horizontal quarter-handrail wreath—transition piece
     (2) Horizontal quarter-string wreath— transition piece

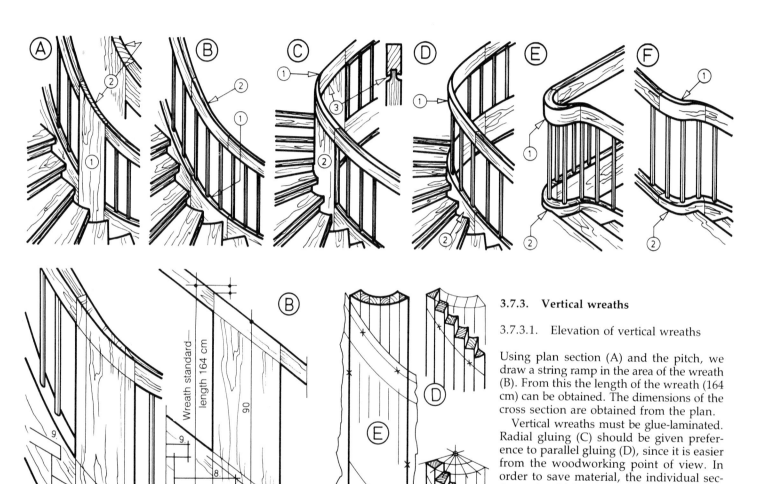

3.7.3. Vertical wreaths

3.7.3.1. Elevation of vertical wreaths

Using plan section (A) and the pitch, we draw a string ramp in the area of the wreath (B). From this the length of the wreath (164 cm) can be obtained. The dimensions of the cross section are obtained from the plan.

Vertical wreaths must be glue-laminated. Radial gluing (C) should be given preference to parallel gluing (D), since it is easier from the woodworking point of view. In order to save material, the individual sections should be staggered in accordance with the pitch of the string.

The paper template shown in section (E) facilitates the exact drawing of the wreath surfaces.

Semicircular wreaths can also be advantageously made by joining two quarter wreaths (F), which should be shaped on the inside (1) before gluing them together.

3.7.4. Horizontal wreaths

3.7.4.1. Design types

Horizontal wreaths can be made as follows:

(A) Solid wood wreath
(B) Glue-laminated wreath
(C) Wreath made from glued segments
(D) Wreath made from core wood and veneer.

(A) Solid wood wreaths are recommended only if they are fairly small and the available wood is completely dry.

(B) Glue-laminated wreaths are the most frequently used types. Semicircular wreaths can be advantageously made by joining two quarter wreaths (1).

(C) Wreaths made from segments joined by gluing and doweling are above all recommended for large diameters, since the grain direction of the individual segments is fairly close to that of the curved string.

(D) Wreaths made from laminated veneer are expensive to make but represent an ideal solution, since they will neither distort nor shrink and provide the highest loadability values.

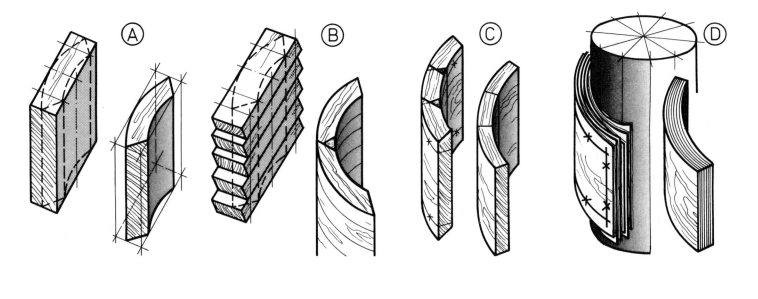

3.7.4.2. Elevation of horizontal quarter-wreaths

The string ramp (A) in the area of the wreaths can subsequently be used for a paper template.

Here we find intersections of lines from the pitch line of the wreath and the plumb lines on the string (1). With the help of the plan shape of the wreath (B), the opening of the elevation (C) can be drawn, with the string ramp (A) being divided by the number of radial lines. The intersections (4) easily provide the opening view (C), to which the length and width of the wood have to be added (5). The thickness of the wood (6) is indicated by the plan. Fig. (D) shows an oblique view of the glue-laminated wreath. Top and bottom of the wood are indicated by a template (7). Its grid pattern is shown by (E).

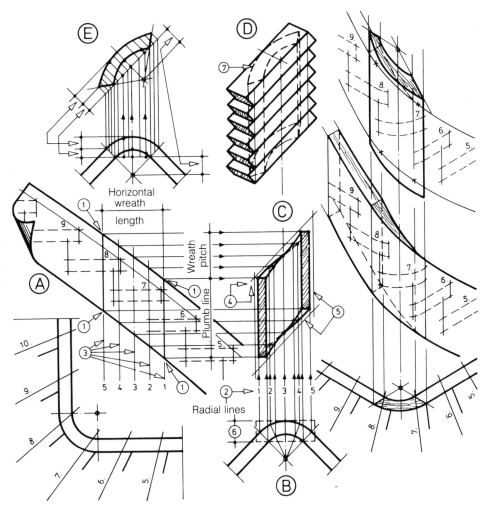

3.7.4.3. Elevation of half-wreaths

As with quarter-wreaths, the following stages apply:

(A) String ramp
(B) Wreath plan
(C) Wreath elevation. Combination of extended elevation (A) and wreath plan (B)
(D) Grid lines determining the template
(E) Indication of the glue-laminated wreath for finishing of exterior surfaces
(F) Wreaths plus strings ready for preparation of the cutouts for the steps.

Fig. (G) shows that the half-wreath represents one half of a hollow cylinder cut at an angle.

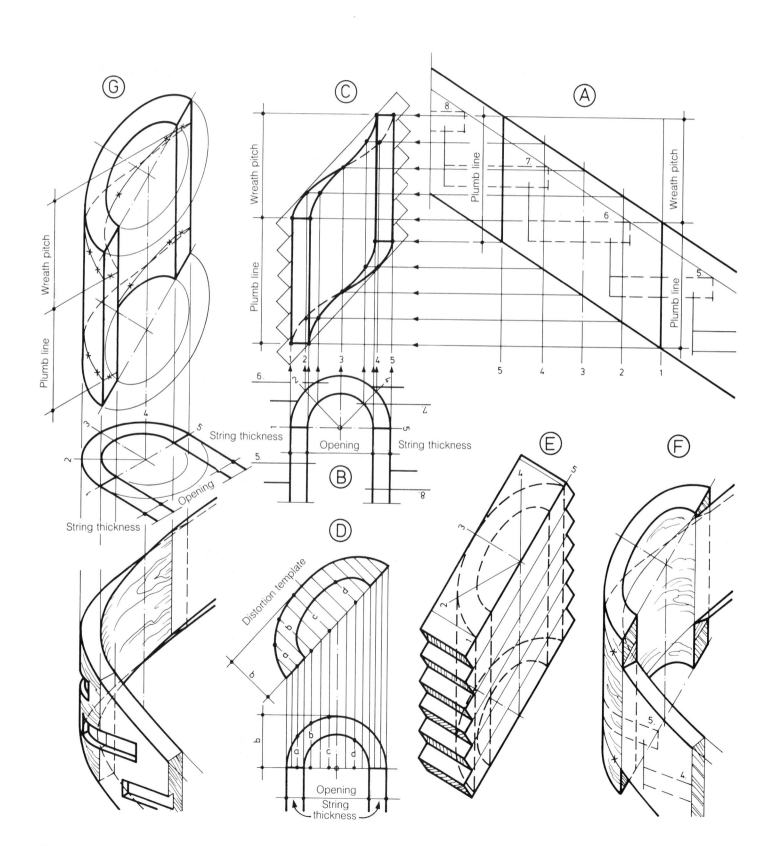

Example (A) shows a staircase with a half-string wreath (1) and a half-handrail wreath (2), both glue-laminated.

For economic reasons, the two wreaths were initially made in one piece and the handrail section (3) cut off the string wreath (4). The cross section (5) shows that the static strength of half-wreaths with a large opening, and in particular those with a shallow pitch, may create problems. With openings above 20 cm, half-wreaths should therefore be made up of two-quarter wreaths (7).

In example (B), two quarter sections have been used to form the transition piece. To this should be added the fact that the grain of the wood flows elegantly over the two sections into the horizontal part of the balustrade, as can be clearly seen from the extended elevation (8).

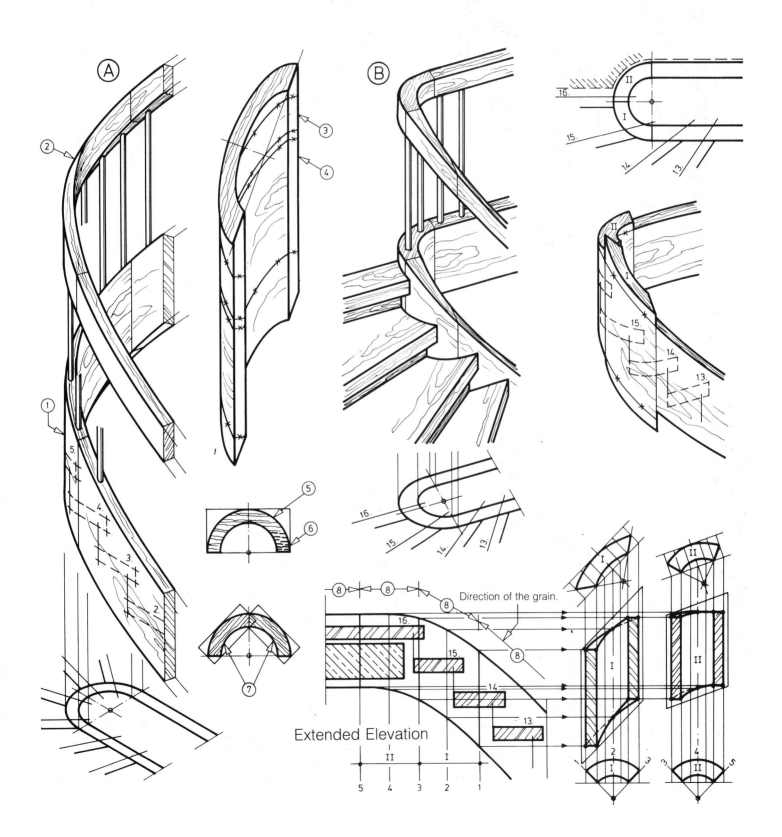

Extended Elevation

Direction of the grain.

43

Half-wreath with elevation of the distortion template:

(A) Consisting of one half arc
(B) Consisting of two quarter arcs (I + II)
(C) Oblique view of the 2 × quarter-wreath before cutting to fit string dimensions

Plan

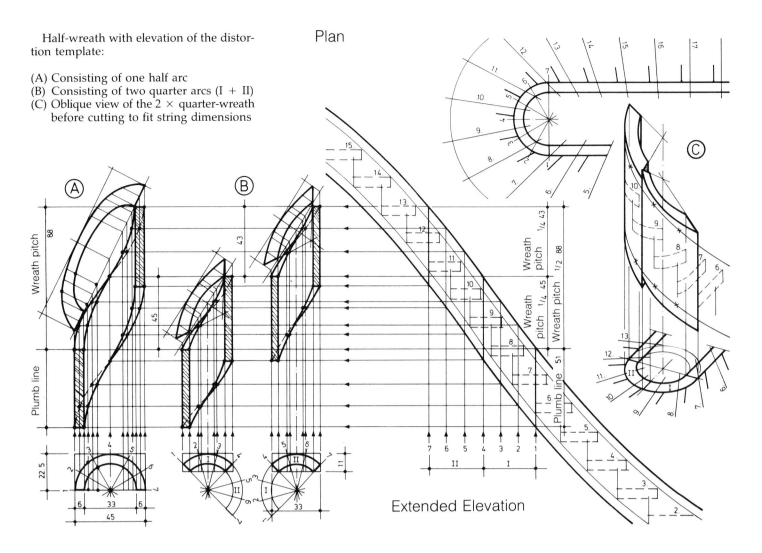

Extended Elevation

Quarter wreath:

(D) Oblique view, schematic
(E) String ramp showing steps

(F) Opening view showing the required wooden planks
(G) Distortion template, elevation through grid lines

Opening view

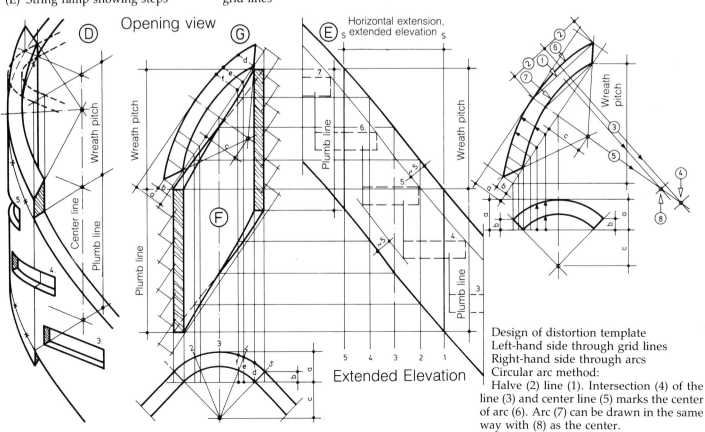

Horizontal extension, extended elevation

Extended Elevation

Design of distortion template
Left-hand side through grid lines
Right-hand side through arcs
Circular arc method:
Halve (2) line (1). Intersection (4) of the line (3) and center line (5) marks the center of arc (6). Arc (7) can be drawn in the same way with (8) as the center.

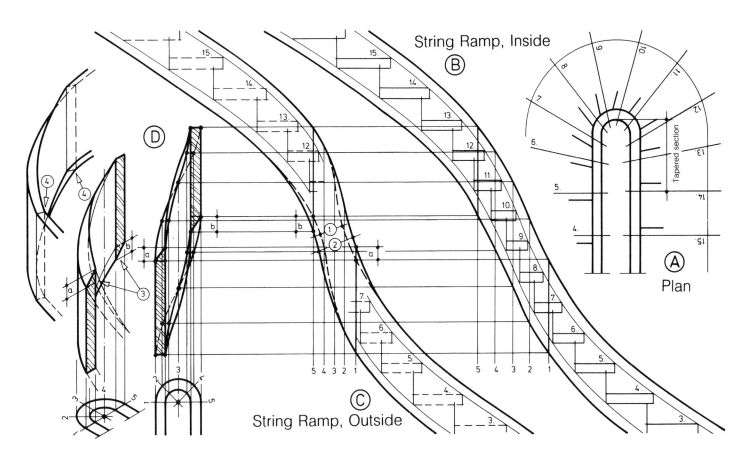

String Ramp, Inside
Ⓑ

Ⓓ

String Ramp, Outside
Ⓒ

Ⓐ Plan

Comparison of wreaths resulting from good and bad tapering of steps.

The plans of the staircases shown above and below are the same, but the tapering is different. In plan (A) the tapered section is too short. The outer elevation (G) is evolved on the basis of the string ramp on the stair side (B), and it becomes apparent that the wreath is too narrow (1). If one adjusts these lines as shown by (2), this changes the plumb-line sections by the dimensions a + b (D)(3). If this modification is not made, unattractive breaks of the curved line result (4).

In the example shown below, the tapered section amounts to approximately one width, plan (E). The tapering is good. A slight modification can be made to the line of the string (F)(1). The line of the wreath is almost ideal (G).

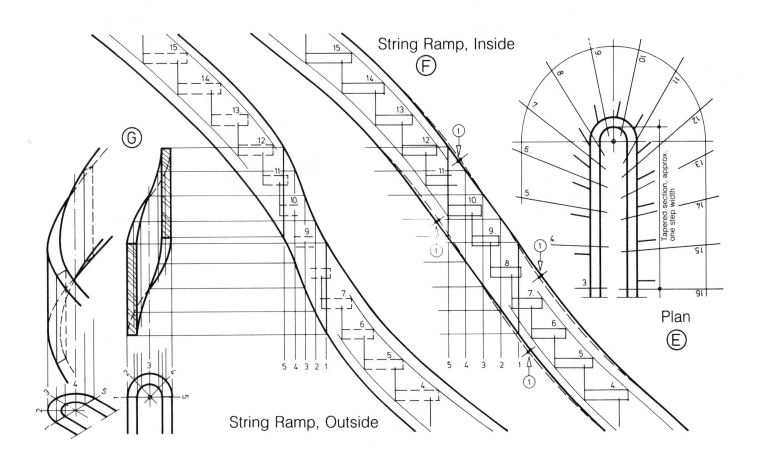

String Ramp, Inside
Ⓕ

Ⓖ

String Ramp, Outside

Plan
Ⓔ

3.8. HISTORIC STAIRCASES—ELEVATION, TAPERING OF STEPS, WREATHS

An old Dutch book published in 1739 by Tieleman van der Horst contains the elevations of staircases and parts of stairs. These drawings of baroque staircases were made at a time when stair construction had reached its zenith, and we recognize with admiration the degree of skill and knowledge that went into their making. One cannot fail to make a comparison with the present day and recognize their challenge. Are we capable of accepting it?

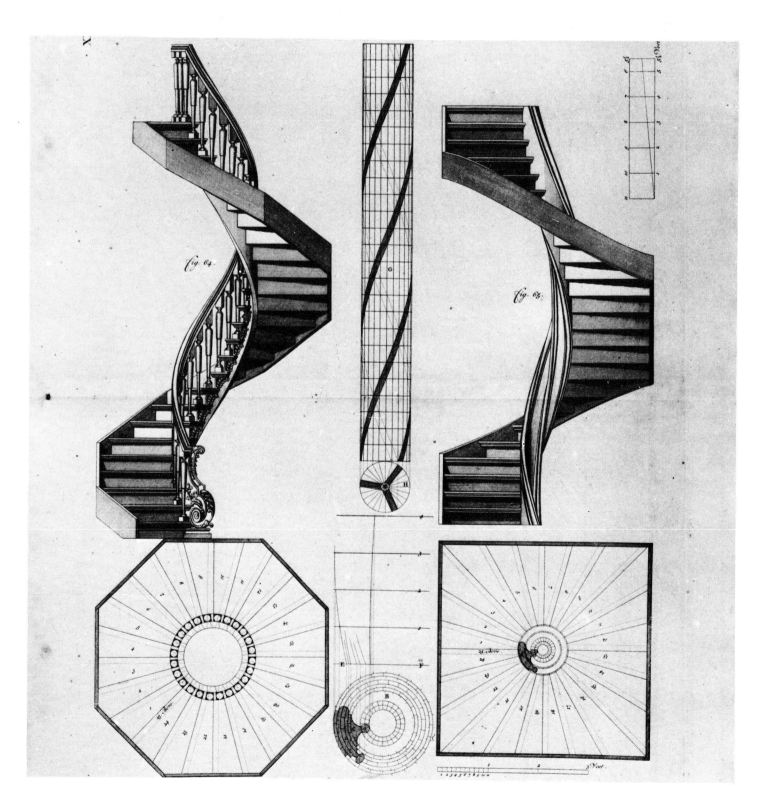

The methods of determining horizontal wreaths through grid lines have remained the same over the centuries. A close examination shows that the wreath camber on the opening side is indicated by the drawing.

The photographs show staircases made in Amsterdam, Holland during the seventeenth and eighteenth centuries.

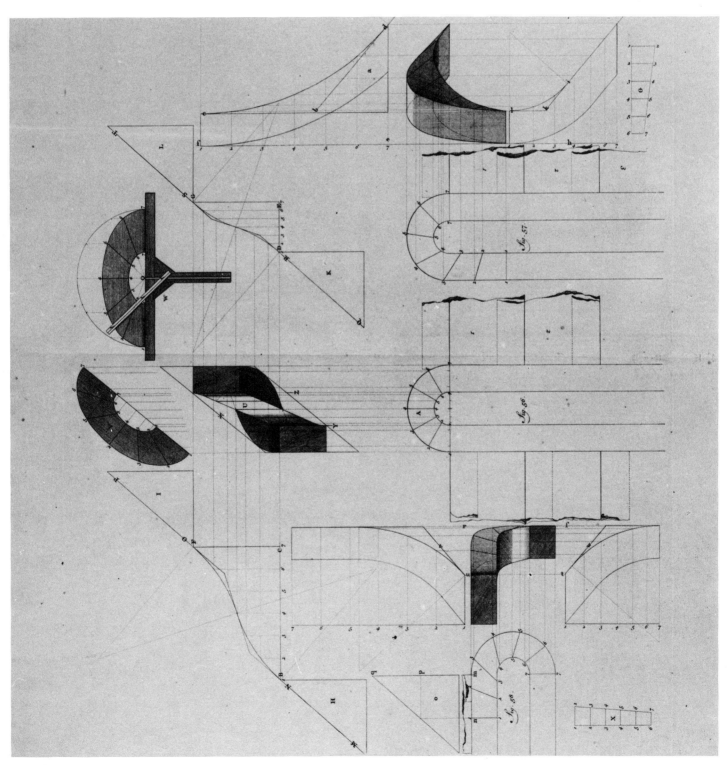

It would be an interesting exercise to carry out static calculations of the saddle-type face string of the stair sections below.

The shape of the molded wreath standard in the bottom right-hand corner of the illustration requires great skill. The flight lines of the carvings on the bannister rods (right-hand picture) are horizontal and parallel to the tread, while their tops are parallel with the handrail.

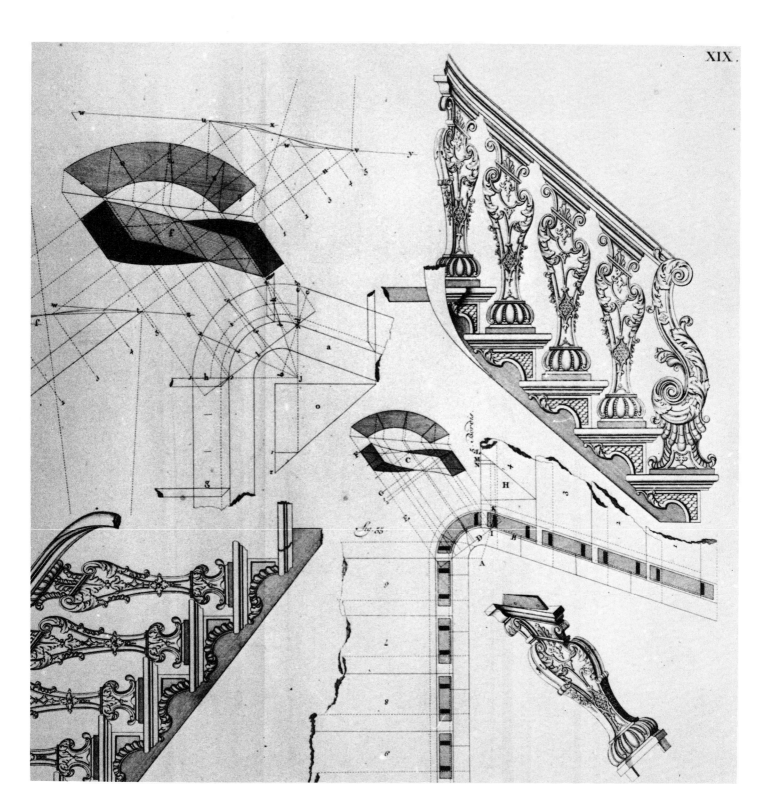

The cross sections of the bannister rods shown below match the edges of the treads and thus become trapezelike.

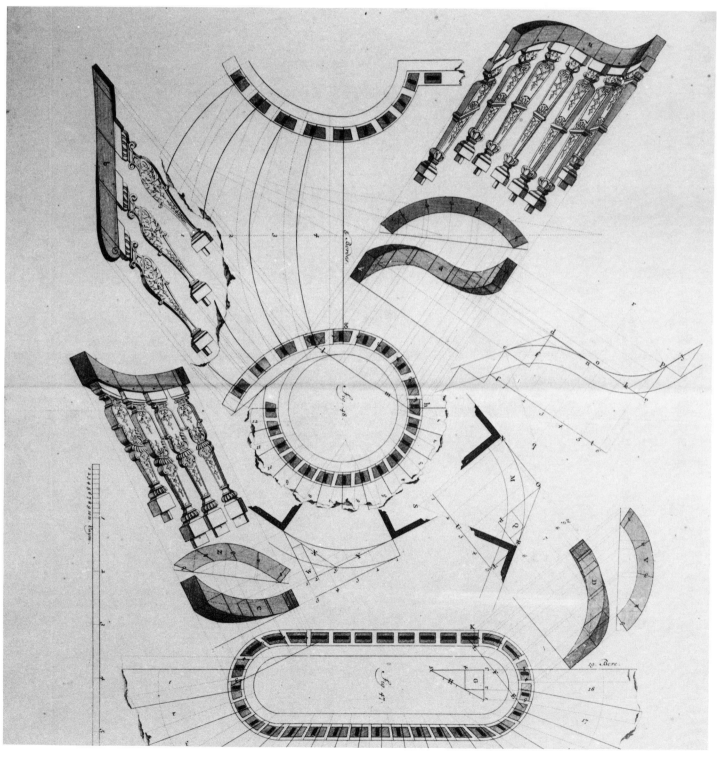

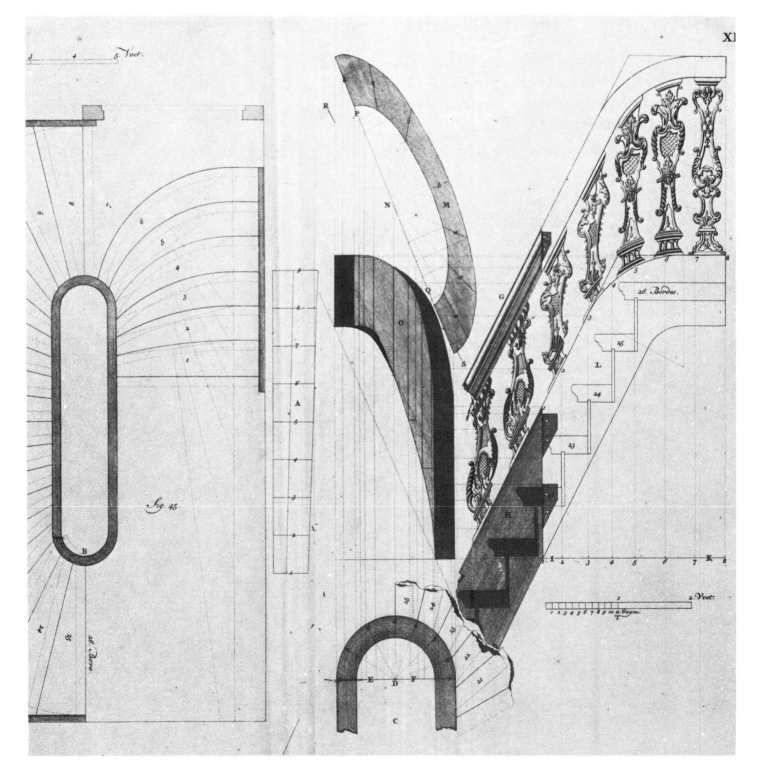

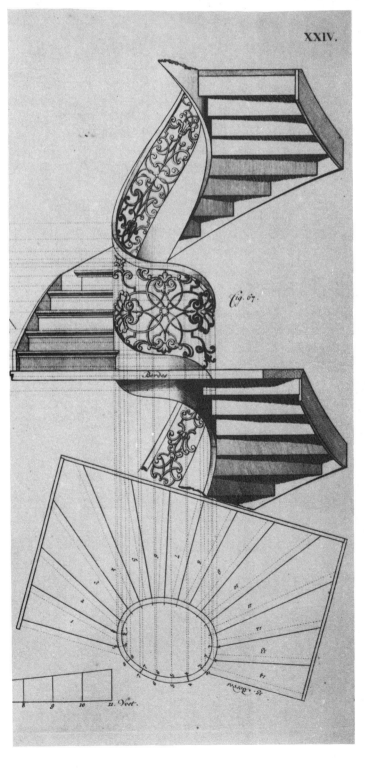

XXIV.

Fig. 67.

Bordes

11. Voet.

15. Bordes

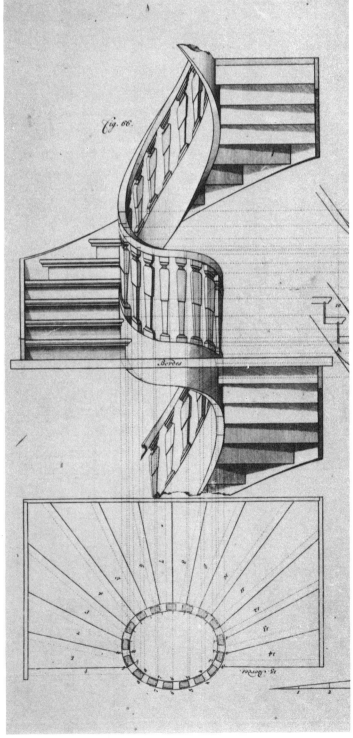

Fig. 66.

Bordes

15. Bordes

3.9. HALF-LANDINGS WITH STAIR RAILINGS ABOVE

3.9.1. Landing types

Landings that form part of a straight flight of stairs are called straight landings (1). With a 90° turn, a quarter-landing (2) is formed; if the flight of stairs turns by 180°, one obtains a half-landing (3).

T = Depth of landing
B = Width of landing

3.9.2. Dimensioning the going of landings in the direction of the walking line

The depth of the landing is to be dimensioned so that the pace length of 63 cm is added to the depth of the tread (A). According to a different rule, the depth of the landing should equal three times the tread depth (B). The latter should be given preference, since it is likely, for example when walking up or down steep steps, that as a result of fatigue one is more likely to take a small step at the landing (B) than one would when negotiating a relatively shallow and more comfortable flight of stairs (C). The dimension of the landing depth should always be shown along the walking line (C). According to DIN 18065, "Staircases in Domestic Buildings, Main Dimensions," a landing should be arranged after a maximum of 18 rises.

There are numerous possible variations on the basic stair shapes, and these should be indicated by relevant drawings.

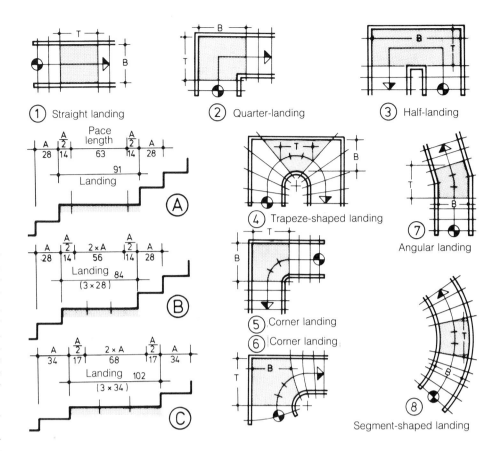

① Straight landing   ② Quarter-landing   ③ Half-landing

④ Trapeze-shaped landing

⑤ Corner landing
⑥ Corner landing

⑦ Angular landing

⑧ Segment-shaped landing

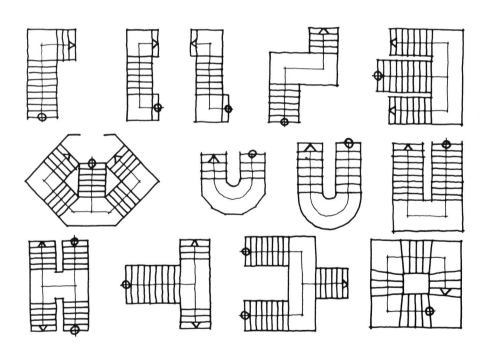

3.9.2.1. Design of quarter-landings at different heights

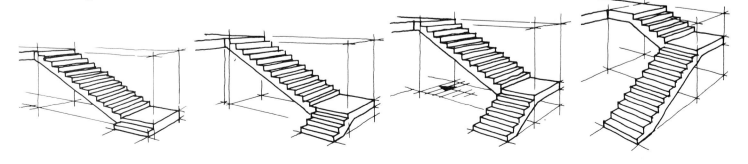

3.9.3. Design of quarter-landings with stair railings above and wreaths

3.9.3.1. Quarter-landings of a string staircase with angular turn of the stair rail

In order to achieve a neat turn in the landing area, the tops of strings and handrails along a plumb line (S) must be on the same level (1). To make this possible, the width of the landing along the string face (string ramp, face side A + B) must equal a total of 28 cm (17 + 11 = 28 cm). Dimensions A + B may differ entirely, but their sum must always add up to the tread dimension. The drawing on the bottom right-hand side shows a similar arrangement but with square turning standards. The optical effect of strings and handrails on one level are shown in sketches (A) and (B).

Such a standard is, for example, needed in period-type staircases with turned bannister rails and standards.

If one compares the space requirements of plans (I), bottom left, and (II), center, one will find that although the tread dimensions are the same, plan (II) has a shorter walking line. Detail B shows that the risers 3 and 4 are practically adjacent. An extended elevation of the face side shows a difference (D) in level between the points where string and handrail standards meet, and these are not acceptable (view I). By tapering steps 3 (landing) and 4 (A) plan (III) or by rounding them along the edges (B) plan (IV), one obtains curved strings and handrails (view II). Although this solution cannot be objected to from the point of view of craftsmanship, one should nevertheless opt for the solution shown in plan (I), despite the fact that it requires slightly more space.

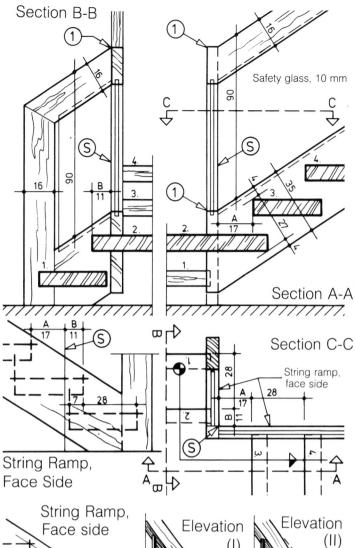

Section B-B

Safety glass, 10 mm

Section A-A

Section C-C

String ramp, face side

String Ramp, Face Side

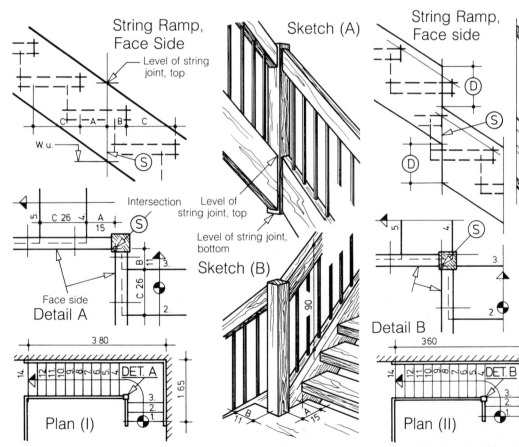

String Ramp, Face Side

Sketch (A)

Level of string joint, top

Intersection

Level of string joint, top

Level of string joint, bottom

Sketch (B)

Face side

Detail A

Plan (I)

String Ramp, Face side

Detail B

Plan (II)

Elevation (I)

Elevation (II)

W.o.

Plan (IV)

Plan (III)

3.9.3.2. Quarter-landings on stairs without strings and with angular turn of the stair rail

With this saddle-type staircase, the arrangement of the landing greatly influences the turning point arrangement of the handrail. The handrail (2) should as a first step be plotted at a distance of approximately 4 cm from the face edge of the steps (1). The internal facing turns around the resulting intersection (S), which ought to be as straight as possible. To demonstrate this point, the facing is shown in the form of a strip of paper (3).

The two dimensions required for the landing (11 + 15 cm) correspond, together, to the subsequent stair depth of 26 cm. Only when this rule is adhered to do we obtain a pleasing handrail turning point (4). The joint appears smoother if the two handrail sections are mitered and the corners rounded (5).

In the drawing shown at the bottom right-hand side, this rule has not been observed. The landing has been designed without taking into account the resulting flow of the handrail: with landing dimensions (E) in the plan of 4 + 4 = 8 cm, measured in each case from the plumb line, it does not add up to the required dimension of 26 cm.

As a result, a difference in level (D) occurs at the plumb line, which makes the handrail (1) inelegant. By producing a flush line at this handrail transition point (2) through mitering, a significant improvement can be made.

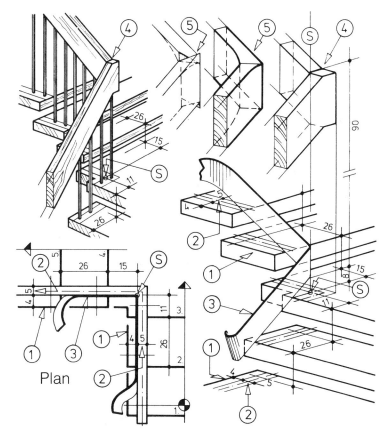

Plan

Example relating to page 53, top right-hand side

(S) = Plumb line (line indicating the intersections of the outer handrail surfaces)
(D) = Difference in the levels of handrails meeting at (S)

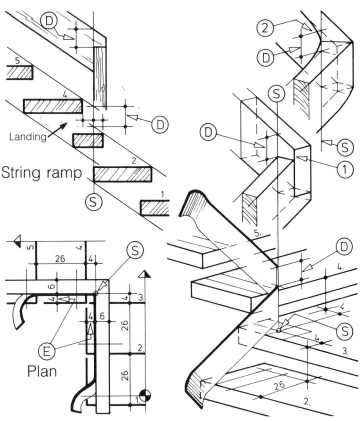

String ramp

Landing

Plan

For the landing arrangement shown in the drawing on the right, two possible handrail solutions offer themselves. If the depth of the landing is not more than about 50 cm, solution 1 is recommended; above 50 cm the horizontal part of the handrail should be brought down to the standard handrail level of 90 cm, and this means solution 2.

In the drawing shown below, the landing has been extended in the other direction. If the extension amounts to less than 50 cm, solution 1 should be applied; if it is more, preference should be given to solution 2.

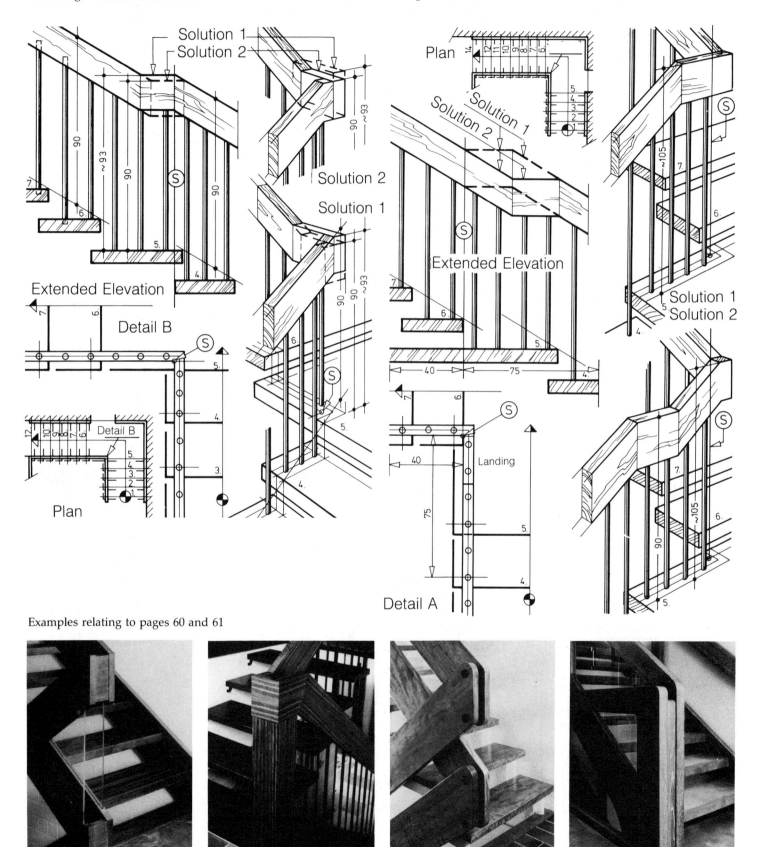

Examples relating to pages 60 and 61

3.9.3.3. Quarter-landings with handrail wreaths for stairs without strings

The connection between handrail sections in the area of the turning points on the landing can best be achieved by the installation of handrail wreaths. The radius on the inside of a quarter-wreath should not be less than 10 cm and need not be more than 20 cm (1), plan detail (B). The face ends of the steps project between 3 and 5 cm beyond the outer edge of the handrail (2). In order to obtain a straight line for the center of the handrail (3) or only a slight bend, the landing dimension along this line (4) must be the same as the stair depth (5). To illustrate this point, the oblique elevation (C) shows a plywood sheet bent above the stairs to indicate the width of the handrail, while fig. (D) shows the thickness of the string on either side of this curved sheet.

In order to determine the grid pattern of the wreath, which is to be designed as a glue-laminated wreath (E), one draws its extended elevation (A). With a horizontal arc length of 21.2 cm, the wreath pitch (KR-ST) is 12 cm. (The resulting pitch triangle (6) has also been incorporated in figs. (C) and (D).)

The wreath grid pattern is applied to the right-hand side of the extended elevation, so that the plumb line intersections (7) on the extended elevation (A) can be linked horizontally with the plumb lines (8) from the wreath plan (E).

The so-called distortion template (10) is also required when making the wreath (F).

The drawing on the right-hand side shows, in an abstract form but nevertheless clearly, that a quarter-wreath is comparable to a quarter section of an obliquely cut pipe.

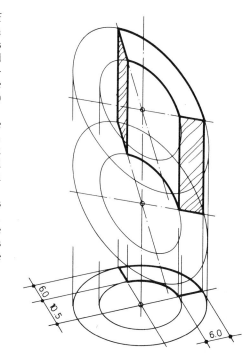

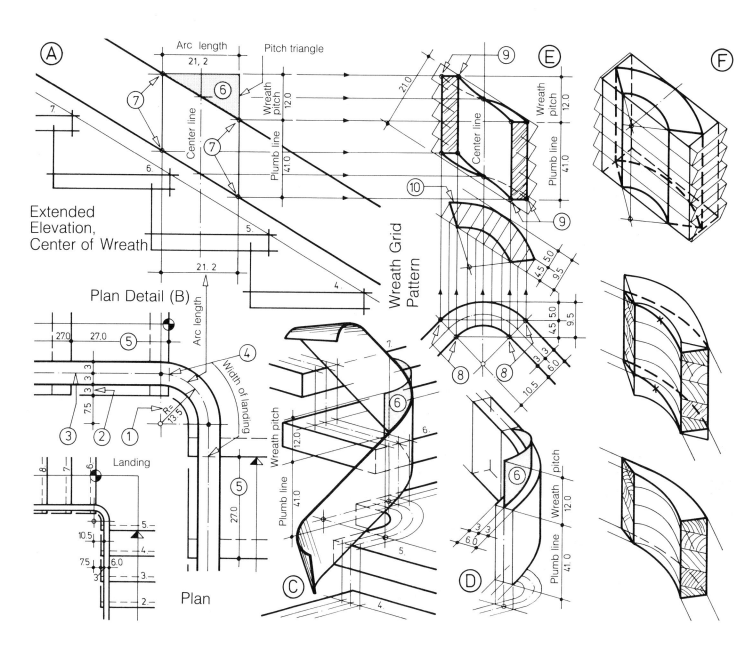

The width of the landing shown in plan detail (A) below, which measures 13.5 cm in the center of the wreath (1), is too narrow. As a result, a marked bend occurs along the string ramp (B)(2), and the wreath becomes too narrow (3). This is even worse when one looks at it from the outside of the wreath (C), since the length of the arc is only 13.35 cm (4). With clever craftsmanship applied to the respective joints it is possible, in such cases, to improve matters by leaving wood at the narrower sections (5) and removing it at the wider parts (6). In order to avoid the resulting problems, the extended wreath should, in such cases, be glue-laminated (E). The quarter arc (7) is extended on both sides by approximately 7 cm (8),

so that all critical sections lie within the wreath. In fig. (F) we see once again how a wreath, badly distorted by a badly thought-out arrangement of steps, can be improved by the above-mentioned corrections (8). In fig. (G), line (9) marks the flight of the descending handrail. The height of 8 cm (10) must therefore be taken care of within the wreath. This dimension can also be found in facing (B)(10).

In the drawing shown in the top right-hand corner, it has been indicated how this glue-laminated wreath can be made up of veneer strips arranged around a central core. The drawing also shows, once again, the straight line of the upper part of the handrail (10).

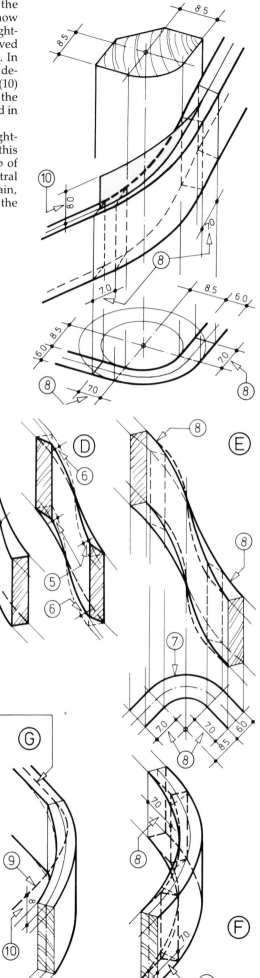

Length of arc, center 18.05

Length of arc, outer 13.5

String Ramp (B), Center of Wreath

Plan Detail (A)

Length of arc
Outer = 13.35 cm
Center = 18.05 cm
Inner = 22.76 cm
Landing width

Plan

Pitch line

Wreath pitch

3.9.4. Half-landings with stair railings above and wreaths

3.9.4.1. Arrangement of steps—stair railings

When designing landing steps, special care must be taken in the arrangement of steps in the area of the turning points. The solutions shown on the right have been subdivided into designs A, B, and C. As can be seen in the plans, the edges of the first and last steps at the turning points have been staggered by one depth in A and C but left aligned in B. Generally, solution B is preferred for reasons that will be discussed below. In all cross sections, system lines have been indicated that limit the length and height of the stair railings; (T) = Stair pitch, (G) = Handrail pitch.

The vertical lines (S) indicate the intersections (1) that normally limit the handrail. In solution B, this length can easily be reduced at the turning points by between 10 and 15 cm so that, compared with examples A and C, the railing and thus the entire staircase can be shortened (cross section B). As a result of this shortening of the handrails, the elevation at which the handrails cross (cross sections B, D) differ in height so that the turning point of the railing will require special attention. Suggested solutions are shown in the subsequent illustrations.

In groups 1–4 the effects of different stair arrangements are shown.

In group 1, angle steps made from artificial stone are placed on reinforced concrete stairs. Here (B-1), the points where the pitch line (3) and the horizontal lower edge of the ceiling below (4) intersect (5) should all be aligned. In solution (B-1), this has been accomplished. This solution may, of course, be impeded by different floor depths. It is, however, possible to stagger these lines slightly (A-1, 6).

Among the solutions of group 2, which shows supporting beam structures made from steel or wood, solution (A-2) should be favored over other solutions because of the finish of the underside. Here the lower edges of the beams meet the face edge of the floor at approximately the same level (A-2, 7). Both beams can be suspended from the ceiling relatively easily by strong steel angles.

In addition to optical benefits, this example has design advantages. In group 3, examples with wooden strings are shown, and here design (B-3) should be aimed for. The lower edges of the strings meet the edge of the floor at roughly the same level (8) and can be attached to the top of the floor. The landing extends over the entire width of the stairwell and optically links the upper and lower stair flights.

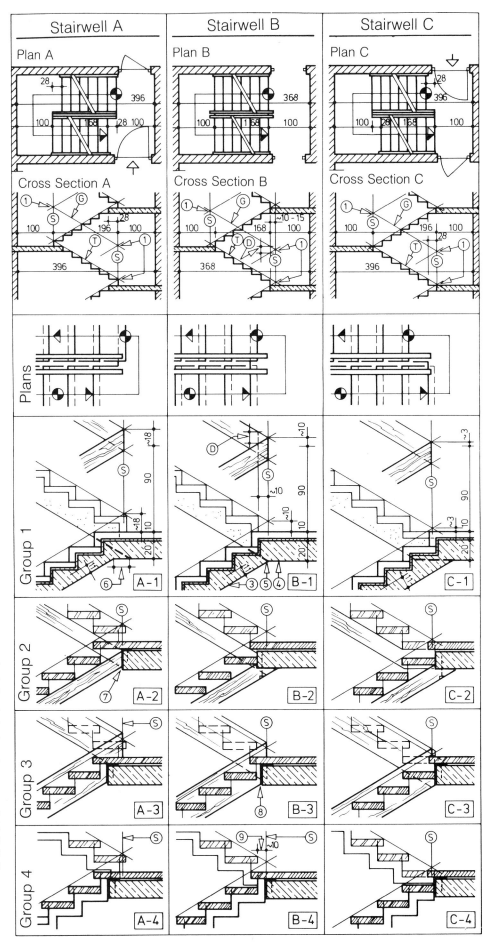

58

Group 4 shows possible solutions for staircases with angular and welded hollow steel beams and artificial stone or wooden treads screwed onto these. Solutions (A-4) and (B-4) are virtually equivalent, although in solution (B-4) the handrail can, if necessary, be shortened by about 10 cm (9).

3.9.4.2. Turning points of angular railings—transition points

The illustration to the right indicates how the handrailing may be shortened by about 10 cm in the direction of the stairs (1)-(D). At the edges (2), a difference in height of about 12 cm (3) results, which is unattractive. The illustration below shows example (A-2) from the previous page and proves, through sketches (A)–(D), that staircase A provides the best solution when supporting beams are being used. In example (A), the hollow steel beams are suspended from the ceiling by steel angles (1). The glue-laminated wooden support beams (3) in example (B) cantilever beyond the box-shaped steel pipe.

In example (C), mitered and welded steel piping (6) projects into the facing wall of the stairwell (4).

The supporting beams and landing beams in example (D) are made from glue-laminated wood (5) and are neatly joined by dowels and screws.

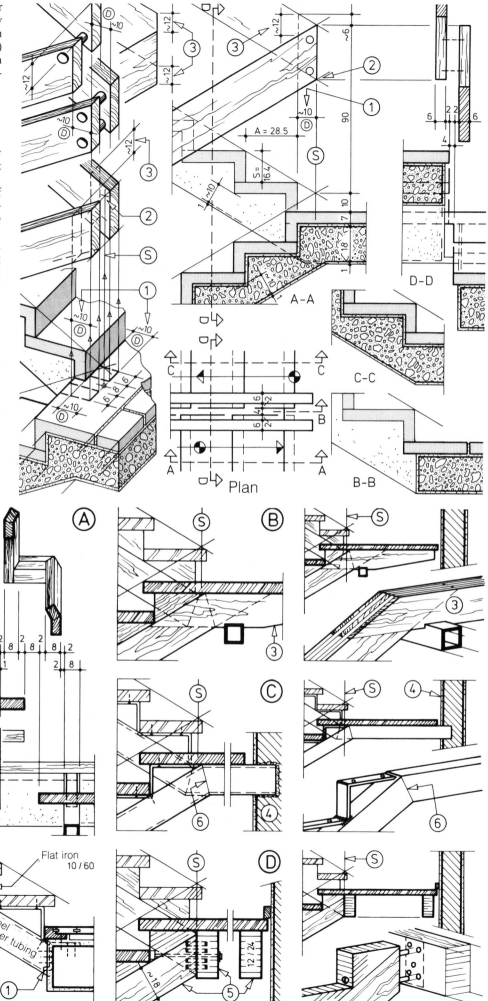

Supporting beam constructions made from bent and welded rectangular tubing are particularly suitable for staircases with landings. Effective spans of up to 2.50 m are possible with relatively small-diameter piping. Supports can easily be provided by a steel angle welded onto the pipe (1). The lower base point should be supported by the ceiling (2). The stair railing, made from flat steel with a PVC handrail, should not form an angle immediately on the plumb line (S)(3), but should be extended horizontally by about 2–3 cm. (4).

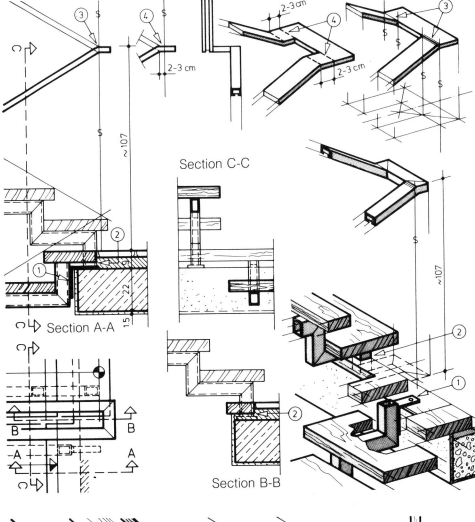

The string staircase (mortised, with open risers) shown on the right, is attached to the landing headers (5), and, in addition, is anchored by steel angles. Projecting string sections can either remain open (6) or be covered by a facing strip (7). If the latter solution is chosen, a similar arrangement will have to be made along the handrail (8). Instead of the two individual edging units, a standard, extending from the top of the floor to the upper edge of the handrail (9), could be inserted. Sketches (10) and (11) show alternative methods of connecting handrails.

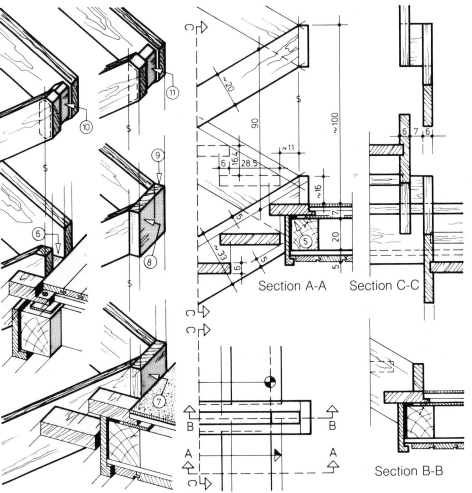

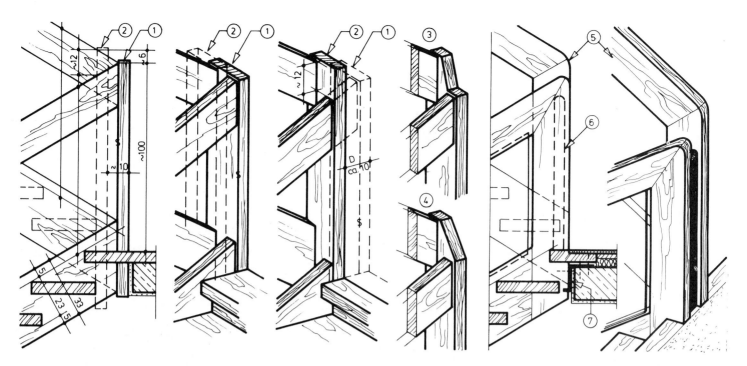

The illustrations above show string stairs with standards at the turning points. Detail (1) shows the more elegant solution, detail (2) a space-saving solution. By choosing suitably shaped standards (3) and (4), optical improvements can be achieved.

This solution becomes considerably more pleasing if the standard is arranged not at right angles but parallel to the handrail and string (5), and the opening is covered by an edging strip (6). As was already shown on page 60, the strings can either be attached to the top of the floor or suspended from the ceiling by steel angles (7).

The illustrations below show, in examples (A) and (B), two different solutions to a handrail turning point. In the case of (A), the joint (1) is arranged so that the angle is halved. This has the advantage that the same handrail profile can be used through-

out. This type of joint is ideal for molded handrails, since exact mitering at the joints results in a flush finish. Example (B) requires more time but is architecturally more elegant. The horizontal handrail sections in the area of the plumb line (2) become wider, and their cross section changes (the depth of the inclined handrail of 20 cm becomes 23.2 cm in the horizontal section of our example). These sections would therefore have to be made separately, which causes no problems in the case of the T-section shown. If the handrail shows complex molding, the separate production of these sections may, however, result in difficulties so that the joint shown in example (A), whereby the angle is halved, should always be given preference. The effects of any differences in level (3) are insignificant.

Example (C) shows a turning point where

the angles have been halved at the joints (the plumb-line solution has been indicated by an intermittent line). The same turning point is shown in (D), using plumb-line joints. Depending on the pitch of the staircase, the widths of the handrails vary. This solution is more balanced and should be preferred to solution (C), especially if a midrail is used.

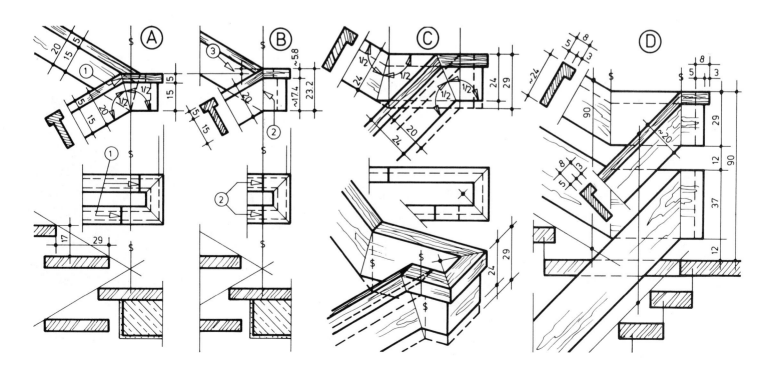

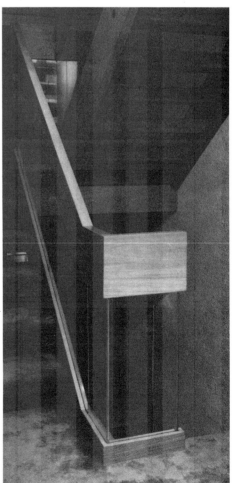

3.9.4.3. Turning points using railing wreaths in staircases without strings

3.9.4.3.1. *Design of a half-wreath between flights of stairs with equal pitch*

In plan (A), the opening between the handrails (stairwell)(1) and the arrangement of steps (2) are shown at a scale of 1:10. The wreath is added later (3). Below this plan, the relevant cross section (B) is shown. The flight line (4) is drawn along the front edges of the steps, and above and parallel to it is the elevation of the handrail (5). One starts the graphic design of the wreath by drawing the string ramp (C) of the stair railing along the center line of the handrail depth (6). The following principle must be borne in mind: the ideal curve of the handrailing is obtained if the pitch of the wreath shown in the extended elevation is, along the center line of the handrail, in line with the pitch of the subsequent handrails or, if the pitch varies, where subsequent handrails are connected by a smooth arc.

A plumb line is drawn (8) above the intersections (7) (cross section B). One starts designing string ramp (C) by drawing the plumb line (9). From cross section (B), a horizontal line should be drawn from the intersections (7) toward the plumb line. From the resulting intersections (10) the handrail pitch, taken from cross section (B), should be marked toward the bottom-right and top-left-hand side (11). Plan (A) provides the diameter for the wreath, which is 18 cm in the center (12); from this, the horizontal arc length of the semicircle can be calculated. This amounts to 28.26 cm and is plotted toward the left and right (13) of the plumb line (9) on the string ramp (C) and marked as plumb-line projection (14). The intersections (15) are transferred onto cross section (B) where they intersect the handrail edges once again (16). The plumb line across the wreath (17)(S) is immediately transferred to plan (A). From the thus-obtained center (18), the plan is completed by the addition of the wreath (20) and its center line. By applying plumb lines onto the wreath in the form of tangents (21), the lateral view (22) can be drawn. In the extended elevation (C) and the cross section (B), the rise/run triangle of the wreath (23) is shown as a shaded gray area. The height of the riser is therefore 17.00 cm with a horizontal string length of 28.26 cm. This dimension could also have been found by plotting the rise/run triangle on an existing handrail edge in cross section B (5). It is nevertheless advantageous to be familiar with the former of these two methods, since it is indispensable in designs that will be discussed below.

Sketches (D) and (E) further illustrate these methods. In sketch (D), a sheet of veneer has been bent to represent the string ramp of the flight of stairs; if additional strips of veneer are glued onto the initial strip on the inside and outside up to a total depth of 3 cm each, this assembly becomes a glue-laminated wreath (see sketch (E)).

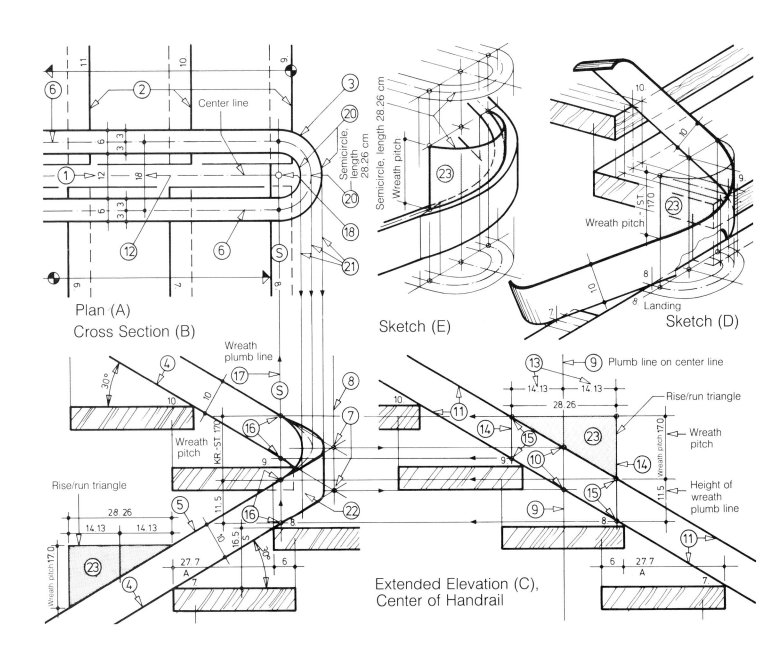

Plan (A)
Cross Section (B)

Sketch (E)

Sketch (D)

Extended Elevation (C),
Center of Handrail

64

This example shows, through three sets of drawings that provide a step-by-step explanation, a method for designing a wreath. In plan (A1) the opening between the handrails (well) (1) and the arrangement of steps (2) is shown at a scale of 1:20. The wreath will be added later (3). Below this plan the cross section (detail) has been drawn (B1), with the flight lines drawn along the front edges of the steps (4) and, parallel to these, the handrail elevations. Where the stair flights have an unequal pitch, one can opt for a plumb-line arrangement of equal height (5), in which case the heights, measured at right angles, differ (6). Alternatively, handrails for flights of stairs with unequal pitch can be of equal height, measured at right angles, in which case the plumb-line dimensions will vary, as will be shown in a subsequent example. The actual graphic design of the wreath starts with the elevation of the so-called extended elevation of the stair railing along the center line of the handrail (C1). In this connection, the following principle should be borne in mind: one obtains an ideal flow of the stair railing across the wreath by aligning the extended elevation in the center of the handrail with the pitch of the adjacent handrails or, in the case of varying pitches, by connecting the adjacent handrails in a soft curve. One draws a main perpendicular line (S) (8) above the intersections (7) and, at an appropriate distance from the former, toward the right of the plumb line (9) on the center line (C1).

From the intersections (7)(B1), horizontal lines are drawn in the direction of the plumb line on the axis. From the resulting intersections (10), the handrail pitch is marked at angles of 30° and 34°, respectively (11). Plan (A2) shows the diameter for the semicircle in the center of the wreath to be 16 cm (12); from this, the horizontal arc length of the semicircle can be calculated. This is 25.2 cm, and one half each of this dimension is plotted to the left and right (13) of the plumb line (sketch C2) in the form of a plumb-line projection (14). The intersections (15) are extended toward the left and intersect the handrail edges (16). The plumb-line dimension (17) can be transferred to plan (A3). From the center (18) thus obtained, the plan is completed by the addition of the wreath (20). By extending the wreath tangents (21)(B3) in the form of a plumb line, the lateral view (22) can be drawn. This view is shown in the bottom left-hand corner (23).

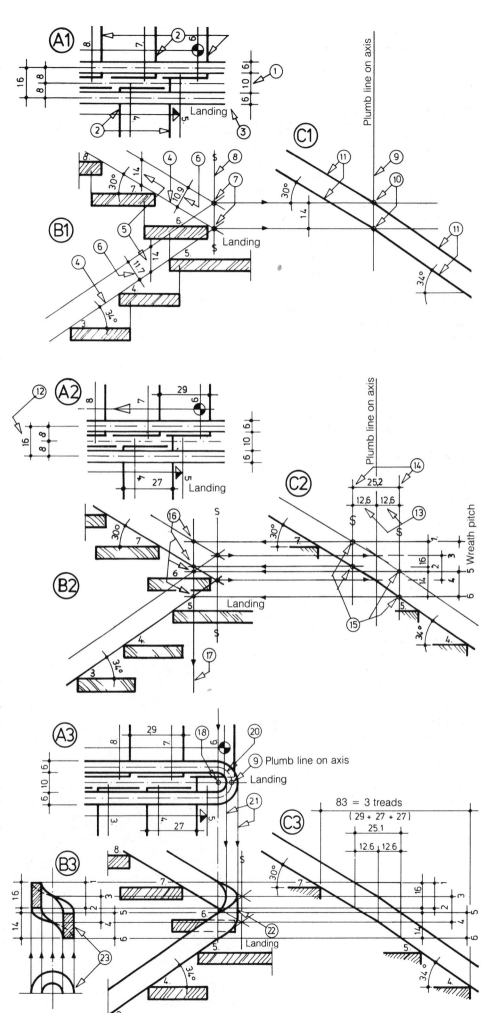

65

3.9.4.4. Turning points on string stairs with stair and railing wreaths

3.9.4.4.1. Design of a half-wreath between flights of stairs with equal pitch

The clear width between the face strings of a flight of stairs with a landing (well) should, as a rule, be between 10 and 20 cm. The arrangement of steps in plan (B) effects the view of the stair flight from underneath (A). An attempt should therefore be made to design the joint with a ceiling so that the lower edge of the string (1) is parallel to the edge of the plaster. In order to achieve this, the top of the lower flight of stairs should be as far back along the edge of the wreath (B2) as possible so that the wreath is at the low point of the concrete ceiling. As has already been discussed in the section on wreaths in stair railings, the center of the stair wreath (3) should be as straight as possible in an extended elevation. In order to achieve this, the dimensions of the steps, as shown in the plan, should be arranged along the center of the wreath and string so that the width of the landing corresponds to that of a step. In order to clarify this point, the central section of the string ramp (D) has been included. Here one plots steps 7 and 8 (4); the string pitch (5) and the plumb line on the wreath (6) are marked using the dimensions taken from plan (B), as 22.5 and 4.5 cm, respectively. One plots the calculated dimension of the extended elevation at the center of the wreath—i.e., 26.0 cm × 3.14 = 81.64 ÷ 2 = 40.82 cm— and this determines the upper plumb line on the wreath (7).

The intersections (8) are transferred to cross section (C), where they form, on the wreath plumb line (9), the points (10), indicating the edges of the upper face string. Parallel to the pitch of the string, a line is drawn across the edges of the steps (11) that provides, at the intersection (12) with the height line of step 11, the front edge of this step (13); this should be transferred to plan (B).

The outer and inner areas of the extended elevation (F) and (G) of the wreath have different pitches as a result of differing basic lengths (E). Since the pitch at the edges of the strings where they meet the wreath (15), as well as the pitch of the strings, always remains the same at the same level, differences in height (16) result, and these must be overcome by slight curving within the respective sections (17).

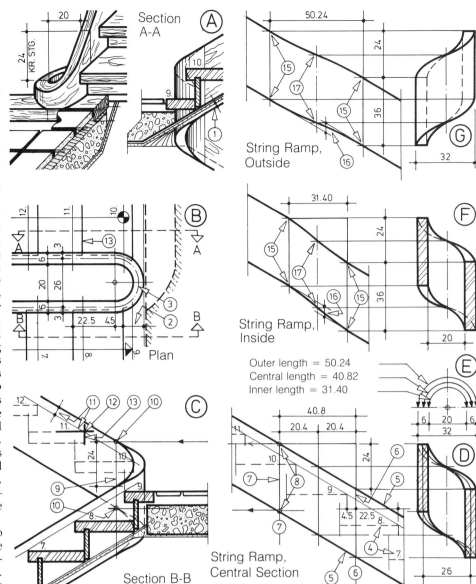

String Ramp, Outside

String Ramp, Inside

Outer length = 50.24
Central length = 40.82
Inner length = 31.40

String Ramp, Central Section

Section A-A

Plan

Section B-B

Design for a flight of stairs extending over two floors, with horizontal quarter-wreath.

3.9.4.4.2. Effects of the arrangement of steps on the view of landings from below

In example (1), the top step 4 (landing) has been staggered by 18 cm (1), compared with the bottom step 5 of the next flight of stairs. As a result, the wreath moves upward at the point where it meets with the ceiling (2), so that the turning point of the handrailing becomes very high. Even worse is the effect on the view of stairs and ceiling from above (3). The points where the two flights of stairs and the ceiling form an angle are so far apart (4) that it becomes difficult to apply the plaster neatly around the wreath. An attempt should be made to keep that part of the string (5) that projects below the plaster visible in the area where it is parallel with the wreaths.

In example (B) these disadvantages have been removed by shifting the positions of the two steps 4 and 5, which are adjacent to the landing. The front edge of step 5 has been staggered by about 8 cm, compared with step 4 (6). This means that the wreath, and consequently the stair railing, are lower at the turning point and thus become more pleasing (7) than was possible in example (A). What is more important, however, is the positive effect on the transition areas of stairs and ceiling (8). The visible string section (9), which projects below the plaster, shows an almost ideal transition from one flight of stairs and the ceiling to the other flight of stairs. In order to obtain a pleasing flow of the string wreath, the edges in pictures (C) and (D) have been incorrectly placed.

The requirement that the width of the landing should, in the rounded section and measured from the center line of the wreath, correspond with the width of the adjacent steps, has not been adhered to.

In example (C), this dimension is approximately 45 cm (10) instead of about 27 cm. As a result, the extended elevation and the view of the turning point show a pitch that is less steep than that of the adjoining sections (11). In example (D), the width of the landing section next to the wreath measures 15 cm and is thus too narrow (12). The wreath pitch becomes too steep (13) and an acceptable curve of the string impossible to achieve. It should be mentioned in this connection that where the wreath has a very small inner radius and a steep pitch, technical problems may arise at the joints with the strings.

Example (C) shows an unfavorable solution; example (D) an altogether poor solution. The extended elevation of the inside (D14) illustrates how the normal string width of about 32 cm is reduced to about 20 cm in the wreath section. The execution of the joints (15) presents technical problems. In order to avoid a poor flow of these lines, the joints can be moved (17) to improve the end result.

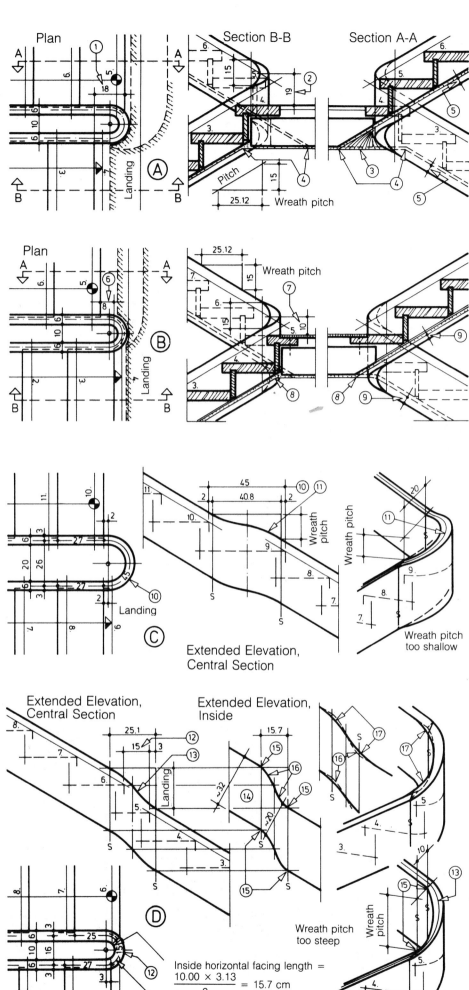

Inside horizontal facing length = $\dfrac{10.00 \times 3.13}{2} = 15.7$ cm

Central horizontal facing length = $\dfrac{16.00 \times 3.14}{2} = 25.12$ cm

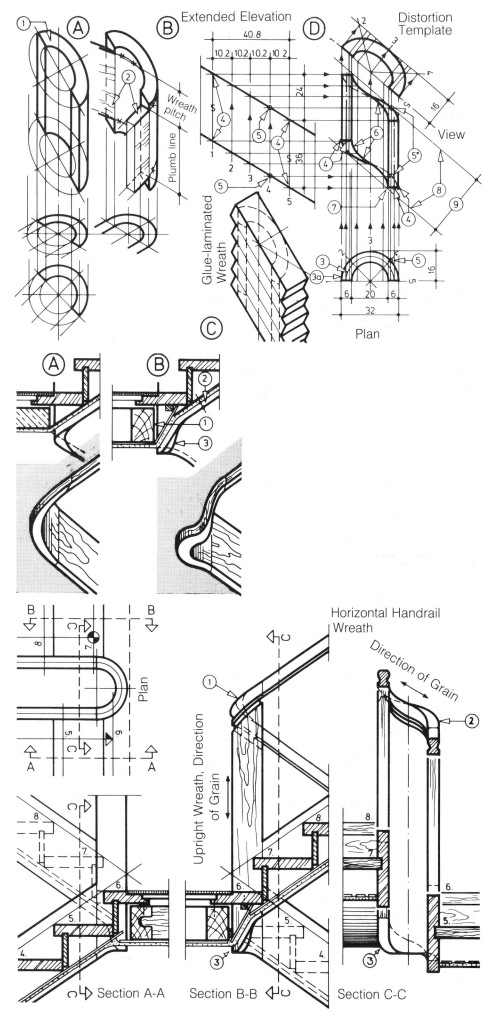

Extended Elevation

Distortion Template

View

Glue-laminated Wreath

Plan

Horizontal Handrail Wreath

Direction of Grain

Upright Wreath, Direction of Grain

Plan

Section A-A   Section B-B   Section C-C

3.9.4.4.3. Grid pattern for the design of a half-landing wreath

For a better understanding of the elevation and production of a semicircular wreath, it should be understood that its shape equals that of a halved, obliquely cut round tube.

This is clearly shown (1) in sketch (A). In sketch (B), the adjacent strings (2) have been added to this "tube section." Fig. (C) shows how, in practice, the wreath is made by glue-laminating several wooden sections. When drawing the grid pattern of the wreath (D), one needs both the plan and the extended elevation. In example (D), use has been made of the central section of the extended elevation (see plan 3). Alternatively, elevation (3A) could be used, since the plumb-line intersections (4) are equally high in both cases. Using the "radial method" (lines 1 to 5), perpendicular lines are drawn upwards (5). The flow lines (6) are plotted above these points (6). Lines (8) are drawn above and below the outer edges (7), and these determine the width of the wood (9). The illustration on the left shows (vertical) wreaths from below.

In fig. (A), the depth of the landing is so small that this view can be designed ideally. In fig. (B), the depth of the floor is greater. For static reasons, it cannot be tapered at the joint with the upper flight of stairs (1). If the plaster is to be applied parallel to the lower edge of the string (2), the wreath will have to be designed so that it includes a "bulge" (3). The lower picture shows the graphic depiction of this procedure.

Vertical wreaths provide maximum stability of the stair-railing turning points in the area of the landing. One the other hand, the flow of the stair railing will not be as elegant as with the previously discussed horizontal wreaths. The molding of the handrail (1) becomes dubious in its upper section, unless a lying handrail wreath is attached (2). A bulge (3) has been added to the lower section of this wreath.

3.10. NEWEL STAIRCASES

The pitch of a newel staircase in the direction of the walking line as shown in fig. (A) (1) is determined by the required headroom (2). The smaller the stair diameter and the narrower the depth of the tread, the steeper the staircase. In order to make a small-diameter newel staircase (up to about 150 cm) as comfortable to negotiate as possible, the walking line should be as long as possible. (The distance of about 37 cm between the center of the handrail and the walking line should not be reduced.) If the front edge of the landing is angular (B)(3), it is possible, despite the small landing dimension along the walking line (4), to obtain a dimension at the top (5) that conforms to the clear width of the stairs. Were the edge at the top straight, the dimension would be too narrow (6). The clear width in fig. (C)(7) is determined on the one hand by the lower edge of the strings or steps (8) and on the other hand by the standard next to step 1 (9). Here too, the width can be increased, as shown in fig. (D)(10), by a rounded bottom step and by taking back the standard (11).

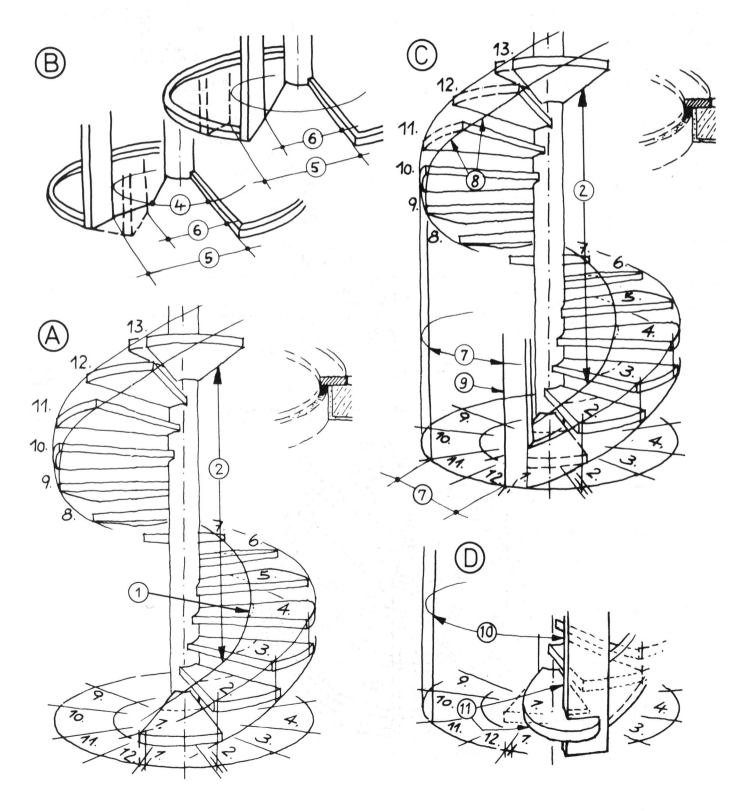

3.10.1. Design of the plan on the basis of the headroom

A newel staircase can be designed on the basis of a plan (A) and the extended elevation of the walking line (B). In plan (A), one first marks the landing, bearing in mind that its dimension in the direction of the walking line (1) should be as small as possible and that the width (2) should be roughly the same as the clear width (3). The rear edge of the landing (4) is transferred to the extended elevation of the walking line (5), after which the headroom of 205 cm (7) at point (8) is marked below the corner of the landing (B)(6).

The length of the walking line is marked at the level of the top landing (72.0 cm × 3.14 = 226.0 cm). The measured point, at a distance of 187 cm (9), is connected with point (8) by an inclined line (pitch line of the walking line)(10).

In order to calculate the rise/run ratio, the resultant basic length of 2.38 m should be divided by the number of rises. Calculation:

Floor height 276 cm ÷ 13 = 21.23 cm per rise

Basic dimension 238 cm ÷ 13 = 18.31 cm per tread

On the basis of these dimensions, the steps can now be marked in the extended elevation (12). We now find that with thirteen rises only twelve treads occur, with the thirteenth step being on the same level as the floor (14). One now spreads the treads over the walking line of plan (A) and tapers the steps accordingly.

In the plan, the front edges of the steps are indicated by a heavy line (15). The center line (16) and the rear edge of the step of the stairs below (17) have been indicated by a short discontinuous line.

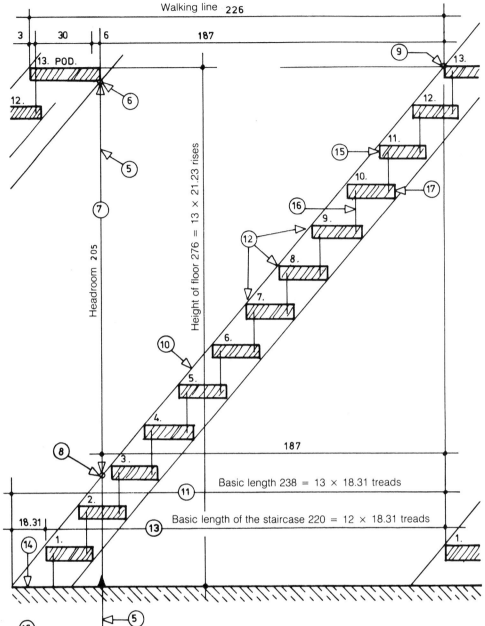

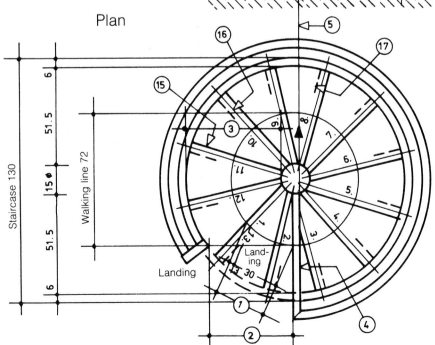

Plan

3.10.2. Starting section of newel staircases

In the case of steep newel staircases, the starting section should first be drawn or checked graphically. This requires that an extended elevation of the outer face of the staircase in this area (A) be drawn. Above the flight line of step 1 in plan (B), the lower (1) and upper (2) elevation sections are projected. The front edge of the lower standard (3) forms the limit on the right-hand side and the line at elevation 1.90 m (4) the limit on the left-hand side. The resulting clear width at the bottom of the stairs of about 68 cm is very narrow. In order to obtain a better width, one could design the staircase even more steeply. The following solution should be given preference: in plan (C), the first step is rounded (5) and the standard has been taken back (6), so that a clear width of about 1 m results. If—as shown in sketch (C)—the standard is turned slightly toward the outside, the opening appears even wider, and the staircase becomes more inviting.

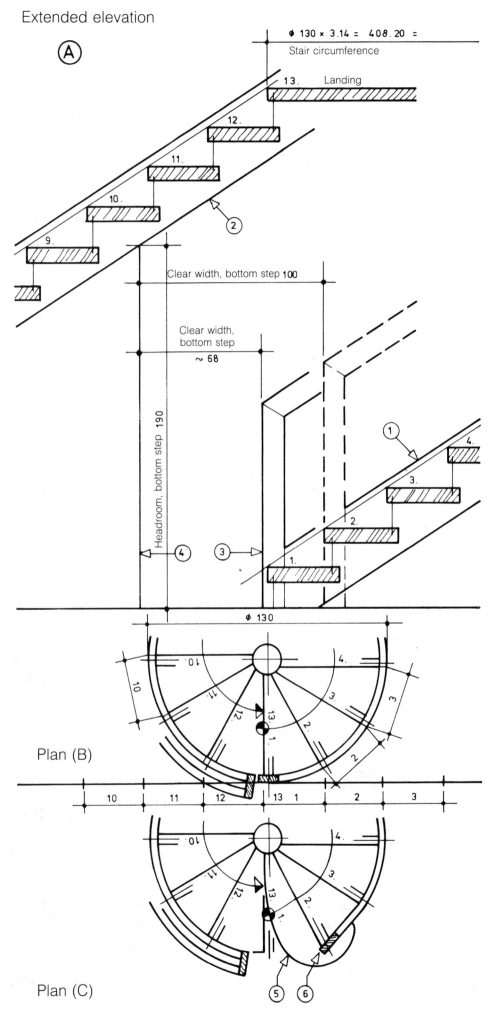

Extended elevation

Plan (B)

Plan (C)

71

3.10.3. Plan, on the basis of headroom and access area, of spiral staircases with a diameter of more than 1.50 m

If the diameter of the spiral staircase is more than about 1.50 m, the top and bottom steps can remain straight since, with increasing diameter, the longer walking line provides a relatively favorable rise/run ratio. In the example shown below, the newel staircase has an outer diameter of 1.80 m. (The same staircase has been shown on the adjacent page. An attempt has been made to find a better solution by providing fifteen instead of fourteen rises.)

Plan (A) has been completed so that the different circles and arcs have been included. In the extended elevation (B), the levels of the various floors (2) and the main axis (3) on the basis of the plan have been marked. The extension of the walking line elevation starts at point (4) in plan (A).

The next step consists of designing the top landing (6) in the extended elevation, which has been moved toward the back by one half of the width of the balustrade handrail (5). Starting at this rear edge, a vertical line, representing the headroom, is drawn (7) and the actual headroom of 2.10 m marked on this line. The length of the walking line (320 cm) marks, at point (8), the front edge of the landing from where, via the lower marking point of the rise/run ratio (9), the pitch of the walking line is marked (10). The basic length of 3.31 m is measured from the intersection of this pitch line with the lower floor.

When calculating the rise/run ratio, the basic length is divided by the number of rises, and the steps can then be plotted.

It now becomes apparent that with the basic length of the staircase, only thirteen treads can be accommodated; the fourteenth is on a level with the floor (11). The access area shown in fig. (C), an extended elevation of the outer edge of the staircase, guarantees a width of 1.24 m and a height of 1.90 m.

Fig. (D) is a schematic drawing showing headroom and access area. If one compared, with hindsight, the rise/run ratio, the width of the landing, the headroom, and the width of the access area, one could, with a further attempt, reduce the headroom to about 2.00 m in order to obtain a shallower pitch of the walking line and thus a more comfortable staircase. It may, for example, be possible to add a further step in order to reduce the rise of 19.86 cm.

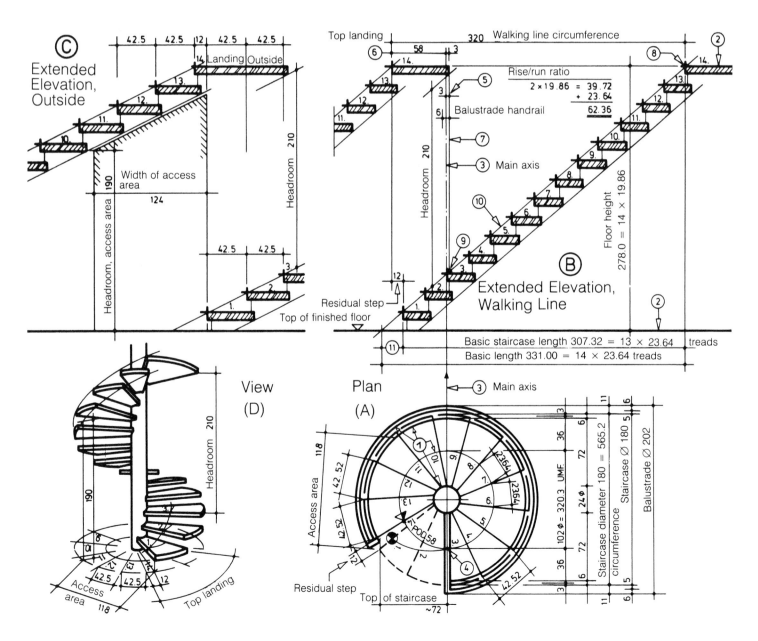

In this example, a further attempt has been made to obtain a shallower pitch line. The headroom (12) in the extended elevation of the walking line (E) has been reduced from 2.10 m to 2.00 m, in order to fit fourteen treads onto the shallower pitch line (13) and the longer basic dimension of the staircase. (Step 15 is now above Step 1.) The design of this staircase has resulted in reduced risers but a less favorable ratio of 59.92 cm (14). An improvement can therefore be obtained only by deeper steps and an extended walking line.

In plan (F), this improvement has been incorporated in the top landing, where a difference of (5 + 25 =) 30 cm (15) is shown. (Difference in walking line between circle and arrow.) This results in an increase in the tread dimension of (30 ÷ 14 =) 2.14 cm, i.e., from 22.86 cm to 25.00 cm, which improves the rise/run ratio from 59.92 to 62.06 cm.

The fact that the access area, as shown in sketch (H), becomes smaller is the price for greater comfort in negotiating the staircase.

In all of these considerations, the position of the staircase, determining whether greater emphasis should be put on the area or the top landing, and whether a pleasing design or practical considerations should be given preference, all have to form part of the final decision.

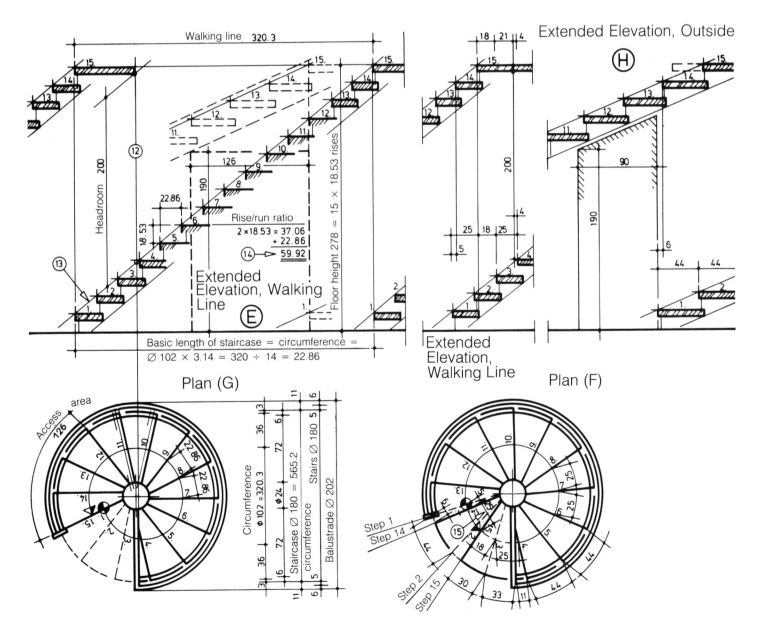

3.10.4. Spiral newel staircases—guide dimensions—types

Spiral staircases should only be used indoors. (DIN 18065, Sheet 1.2.4. stipulates that the tread of spiral staircases must measure at least 100 mm at a distance of 150 mm from the narrowest part of the step.) This regulation does not make sense in the case of spiral staircases. It produces enormous problems and should be modified. The walking line of spiral staircases is arranged at a distance of 35 to 40 cm from the center of the outer handrail.

The plans of typical staircases shown below are provided as design aids.

Example: Existing finished floor height 2.70 m; unfinished stairwell 1.50 m. In the column of staircases 150, we look for the floor-height range which, for our example, is type 150-12. The appropriate plan can be added to the working plan; scale 1:50. This does not replace the drawing up of a work plan where minor modifications of step arrangement may be possible or necessary.

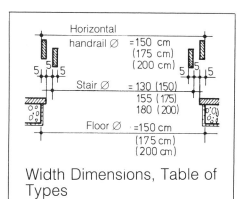

Width Dimensions, Table of Types

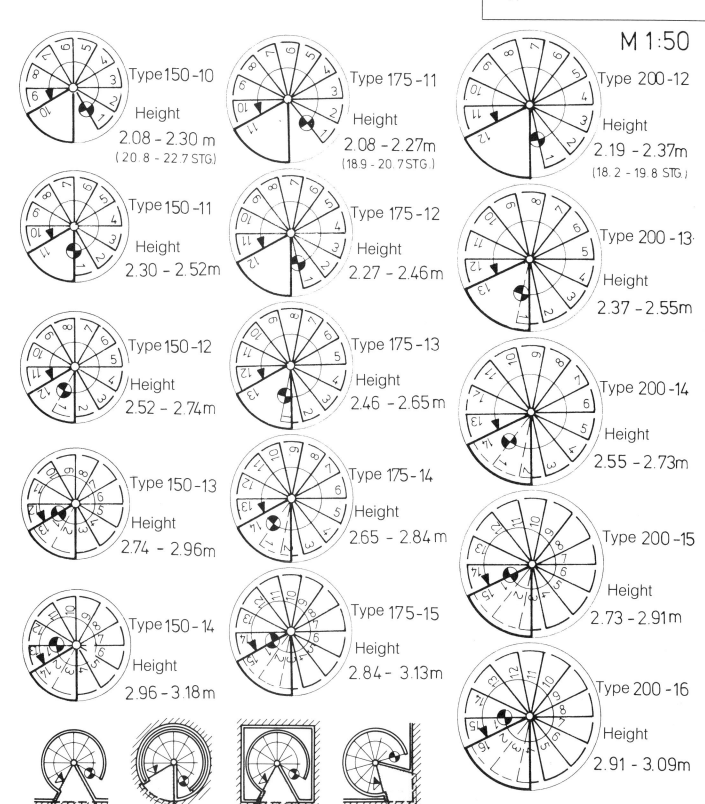

Type 150-10 Height 2.08 - 2.30 m (20.8 - 22.7 STG.)

Type 150-11 Height 2.30 - 2.52 m

Type 150-12 Height 2.52 - 2.74 m

Type 150-13 Height 2.74 - 2.96 m

Type 150-14 Height 2.96 - 3.18 m

Type 175-11 Height 2.08 - 2.27 m (18.9 - 20.7 STG.)

Type 175-12 Height 2.27 - 2.46 m

Type 175-13 Height 2.46 - 2.65 m

Type 175-14 Height 2.65 - 2.84 m

Type 175-15 Height 2.84 - 3.13 m

M 1:50

Type 200-12 Height 2.19 - 2.37 m (18.2 - 19.8 STG.)

Type 200-13 Height 2.37 - 2.55 m

Type 200-14 Height 2.55 - 2.73 m

Type 200-15 Height 2.73 - 2.91 m

Type 200-16 Height 2.91 - 3.09 m

3.10.5. Spiral staircases with unusual plans

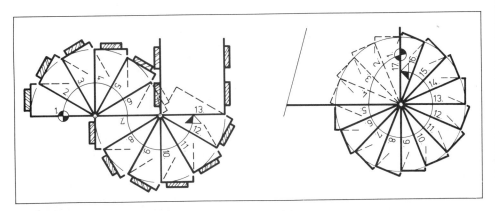

3.10.5.1. Spiral staircases with unusual starting and finishing arrangements

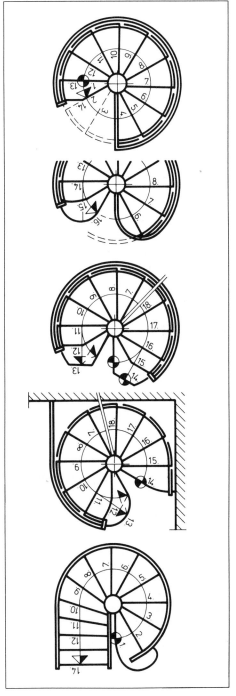

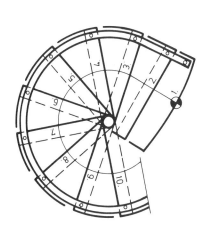

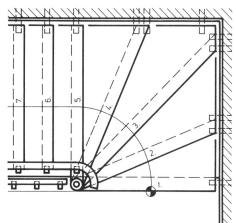

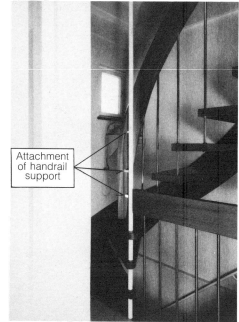

Attachment of handrail support

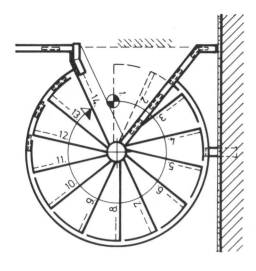

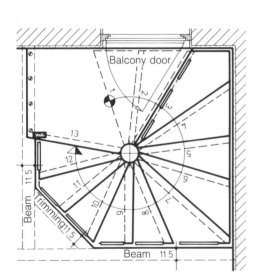

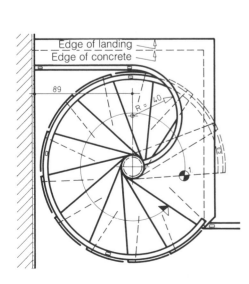

Spiral staircases provide a wide variety of design possibilities, even where they are used as part of an existing setting. The stairs shown on the right lead to a slide that has been boldly placed into its natural surroundings; the actual slide has a functional stainless steel cover but has, in design terms, a subordinate function and has thus become unobtrusive.

The staircase shown below was built in a teacher training college and has all the features of a wooden sculpture. Arranged as a piece of "architectural art" in the corner of the assembly hall, it is frequently used by students, although a comfortable reinforced concrete staircase is in its immediate vicinity.

Spiral staircase with flattened and twisted
newel

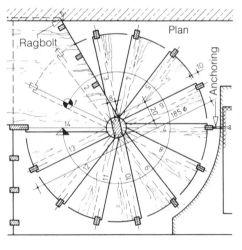

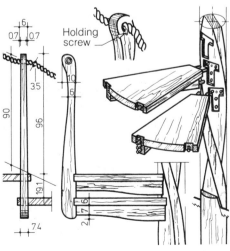

Plan

Ragbolt

Anchoring

185 Φ

Holding
screw

3.11. COVERING STAIRS WITH TEXTILE OR PVC AND WITH STAIR NOSING

PVC or textile covers should always be glued onto the steps. The edges should be well rounded (A) so that no bubbles or hollow areas can form when the cover is drawn across the step (B). In case of overlapping, the tension properties of the adhesive must be borne in mind (C),(D). Where solid wooden steps are used, it must be remembered that subsequent shrinking may occur. If PVC covering is glued onto these steps, bubbles may form. This possibility does not exist with textile covers.

(B) Incorrect

(A) Correct

(C) Correct

(D) Incorrect

Textile Cover with Metal Profiles

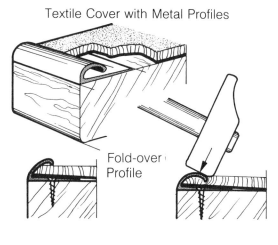

Fold-over Profile

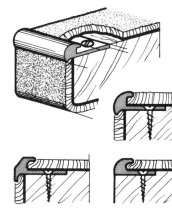

Combination Metal Profiles

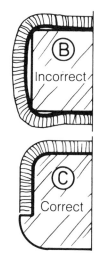

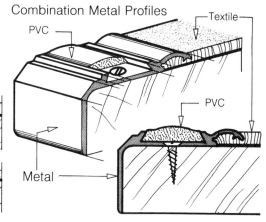

PVC — Textile

PVC

Metal

Textile Cover, Continuous

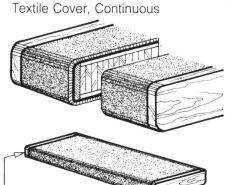

Facing glued on

Textile Cover, Sunk Front and Top

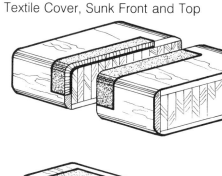

Textile Cover, Continuous, Folded around Facing

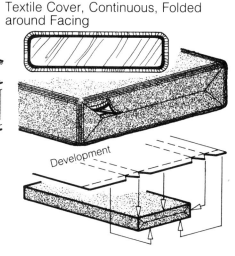

Development

Textile Covers with PVC Edging

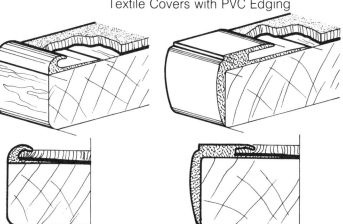

Weld

Artificial Stone Steps with Protective Strips Inserted into the Concrete

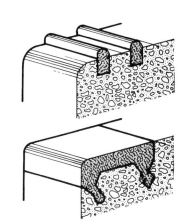

3.11.1 Covering steps and strings with textile materials

In this example, steps recessed into the wall have been covered with carpeting on all sides (1). The joint is on the underside of the step (2). The edges of the steps should be rounded, in line with the bendability of the cover (3).

The same cover has been glued onto the wall area parallel to the pitch of the stairs (4). In order to obtain a neat finish, skirting boards made from rigid PVC have been attached to the unfinished wall (5). These are continued in the form of skirting (6).

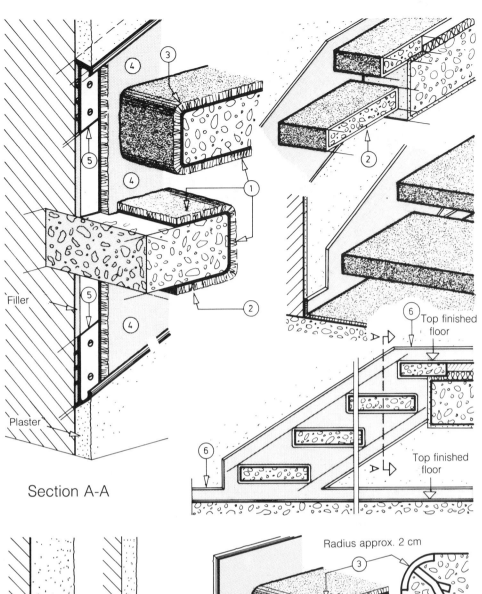

Section A-A

This illustration shows the staircase made from in-situ cast concrete and covered with carpeting; skirting boards in the shape of and parallel to the steps have been fitted to the wall (1). It must be remembered that these skirtings should be continued at floor level (2). In order to obtain neatly rounded edges, quartered piping with attached anchors can be incorporated during casting (3).

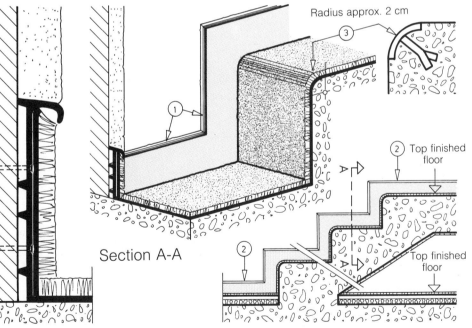

Section A-A

81

3.12. RENOVATING OLD STAIRS

3.12.1. Lining with wood

When applying additional layers of wood, it must be remembered that the first riser will become higher and the last one lower.

The new step shown in fig. (A) is of a plywood-type design and shaped in the form of a wedge. Its surface consists of 6-mm sawn veneer (1). The new step is inserted into the carefully shaped old step (2), glued in front (3), and nailed at the back (4).

The nails are covered by the new riser, which supports the old step above (5).

Attachment of a hardwood edge and a mosaic parquet cover.

Fig. (B): The following stages have to be completed:

(1) Cutting off of the nosing
(2) Preparation of a new nosing strip
(3) Doweling on of the nosing strip
(4) Applying filler to the top surfaces
(5) Attachment, by screwing, of a 6- to 8-mm-thick chipboard or plywood sheet
(6) Laying and gluing on of 9-mm mosaic parquet
(7) Attachment of nosing.

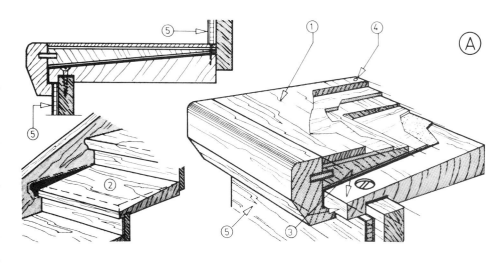

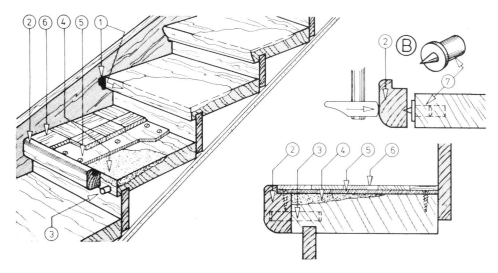

Fig. (C) shows a similar renovation process involving mosaic parquet.

Here, a section of the step has been removed by means of a handtool with an appropriate attachment (10). The other operational steps are the same as those described in fig. (A). In cases where the thickness of the remaining wood is insufficient (11), solution (D) should be chosen.

Additional steps:

Attach riser (12); screw on riser (13).

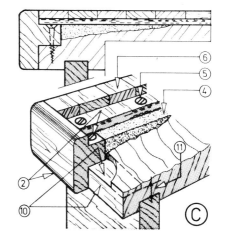

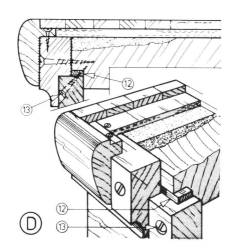

3.12.2. Lining with textiles or PVC

Attachment of stair nosing made from PVC or metal. Example (E):

(1) Cutting off of the old stair nosing
(2) Attachment of the new edge
(3) Leveling with filler
(4) Application of a chipboard or plywood sheet and screwing into place
(5) Gluing on PVC nosing
(6) Applying filler
(7) Gluing on PVC or textile cover.

The stair nosings in fig. (F) are made from nonferrous metals (LM) or brass (MS). In the case of stairs subject to frequent use, it is also recommended that an antislip and wear-resistant insert be used (1).

Application of textile cover, fig. (G):

If the textile cover is to be extended over the front edge of the step, high quality material must be used, and the radius of the wooden edge should be as large as possible.

(1) Screw prepared wooden strip and cover onto step
(2) Glue and screw plywood sheet into position
(3) Glue cover into place.

3.12.3. Creaking stairs—curing the creaking noise

Creaking noises arise when two surfaces rub against each other. In the majority of cases, this is due to a shrinkage in the risers.

Sketch (A) shows the shrunken area (1). When a load is applied to the tread (3), the surfaces (2) rub against each other. In sketch (B) the tread is screwed onto the riser and then covered with a textile cover.

Fig. (E) shows another good solution. The step is "jacked up" by means of a block of wood (1) and a strip of wood is glued and screwed into the corner between riser and tread.

In Example (F) a new riser (2), curved at the top, is inserted.

Creaking noises are caused by a shrinking of the treads—fig. (G). This can be cured simply by screwing the tread onto the string (1) or by inserting and gluing in of wedges (2). Hollow-sounding stairs, as shown in fig. (H), can be filled with a foaming agent (3), which can be poured into the cavity formed by lining (1), lower edge of the step, and rear edge of the riser (2). The holes through which the foaming agent has been poured can then be mended during the course of renovation work (4).

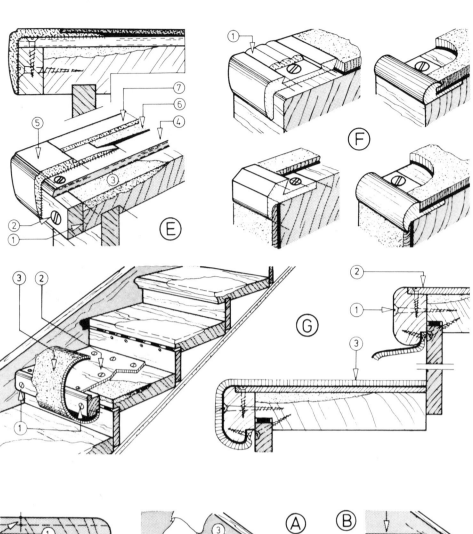

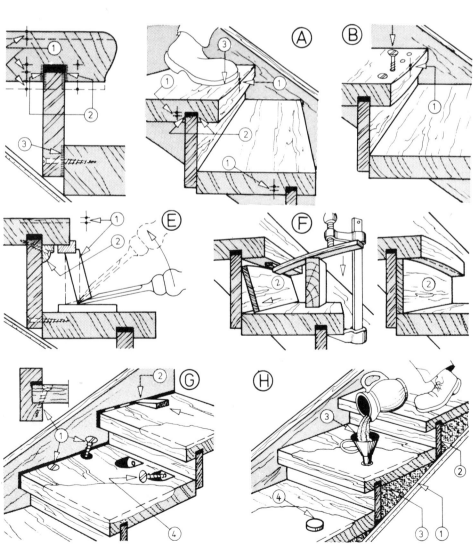

3.13. STAIR RAILINGS

3.13.1. Stair railings made from ornamental ironwork

Wrought ironwork for handrails is commercially available today and lends itself to a great variety of combinations.

Handrail profiles
Bottom chord profiles
(Made from iron, brass, copper, bronze)

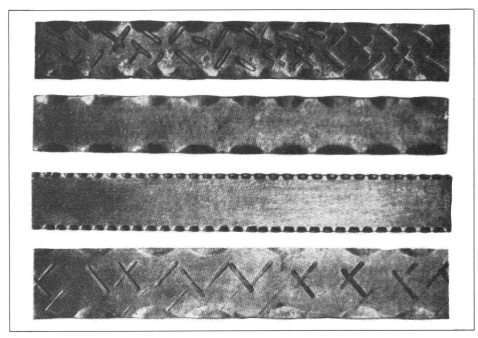

Ornamental ironwork sections, some hammered flat

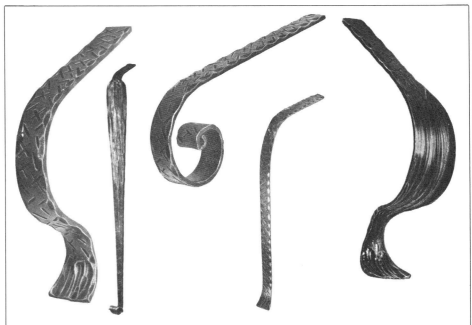

Rosettes for covering square and round steel rods

Bannister rods

Details of ornaments

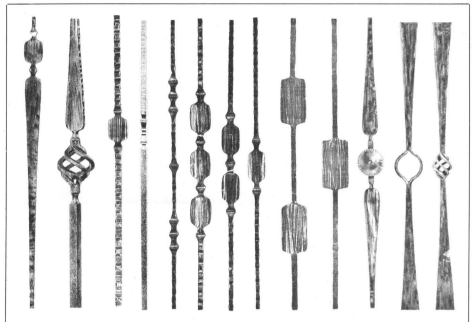

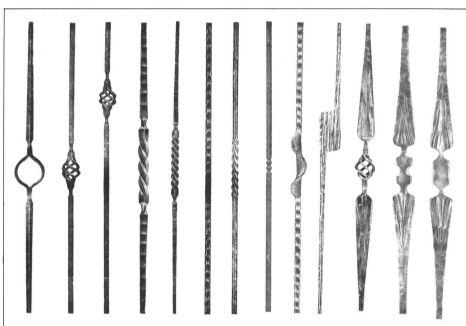

Stair railings made from ornamental ironwork can successfully utilize the many design possibilities that this material offers. By a skillful combination of square or flat cross sections, the railing can become three dimensional when viewed from different angles. The material can be shaped into a wide range of patterns, ranging from the plain to the artistic or ornate.

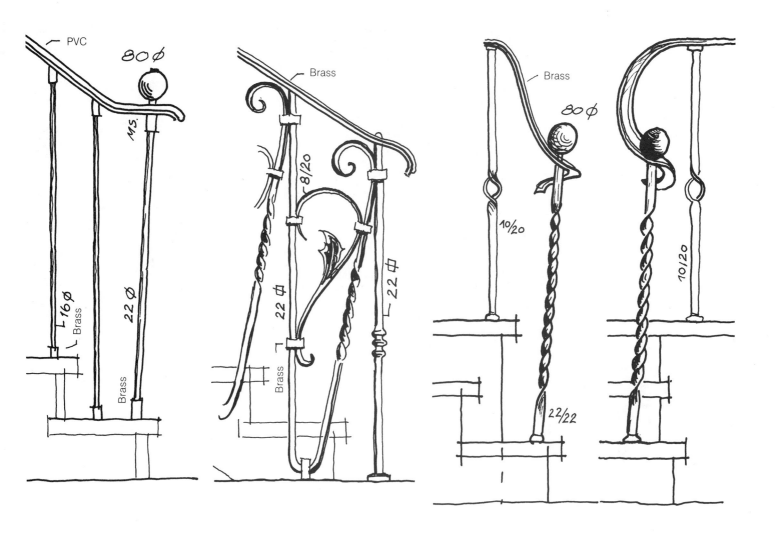

Starting posts

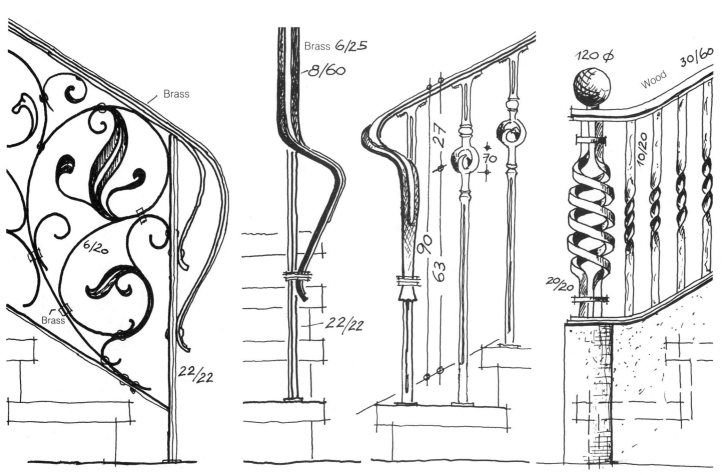

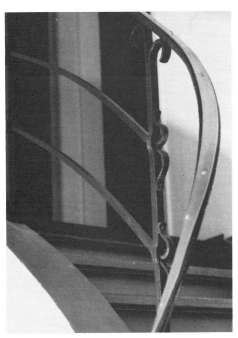
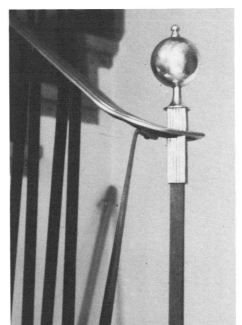

3.13.2. Steel stair railings on steel stairs

Chromium-plated newel staircase

3.13.3. Wooden stair railings

3.13.3.1. Stair railing components

1 and 3	Cylindrical bannister rod
2	Cylindrical bannister rod with ornament attached
4 and 5	Turned bannister rods
6 to 10	Turned bannister rods with circular or square end sections
11 to 14	Rectangular bannister rods
15 to 25	Bannister rods above 5 cm in diameter
26 to 31	Bannister rods up to 5 cm in diameter

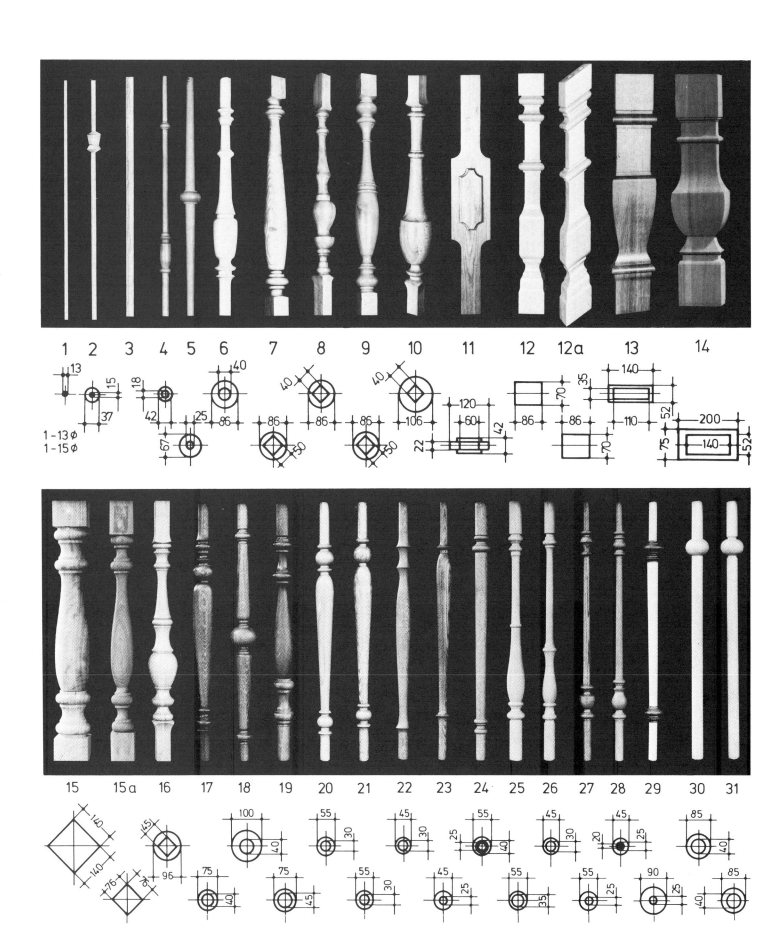

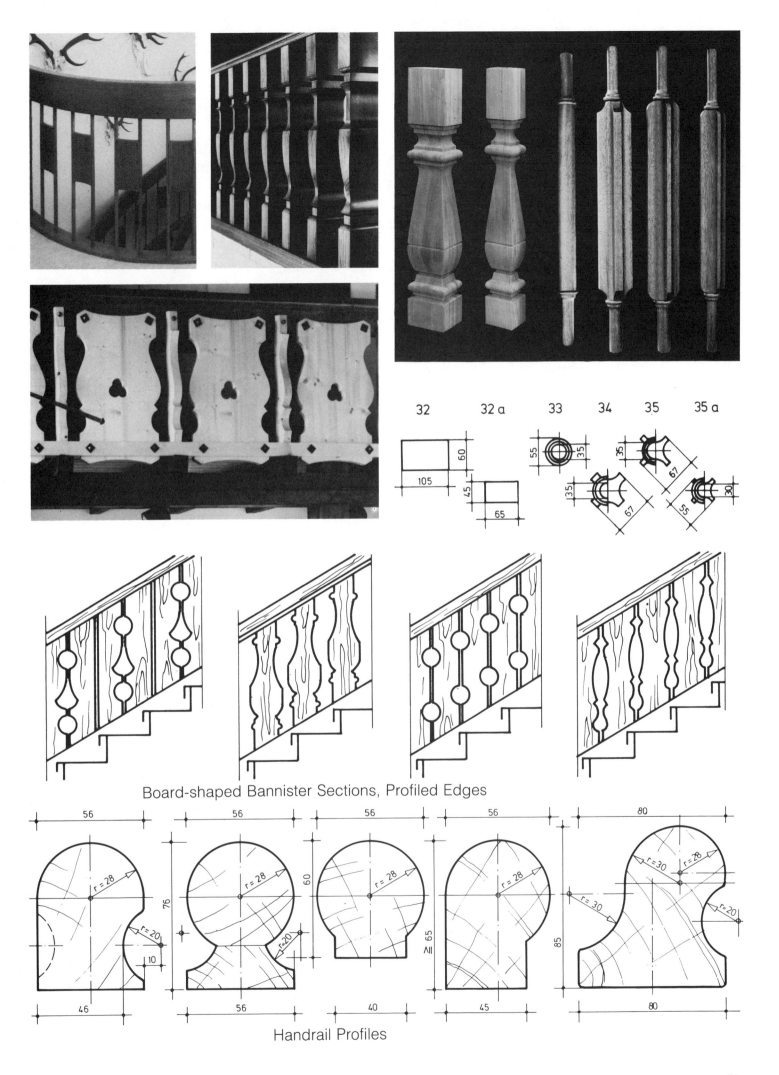

32 32 a 33 34 35 35 a

Board-shaped Bannister Sections, Profiled Edges

Handrail Profiles

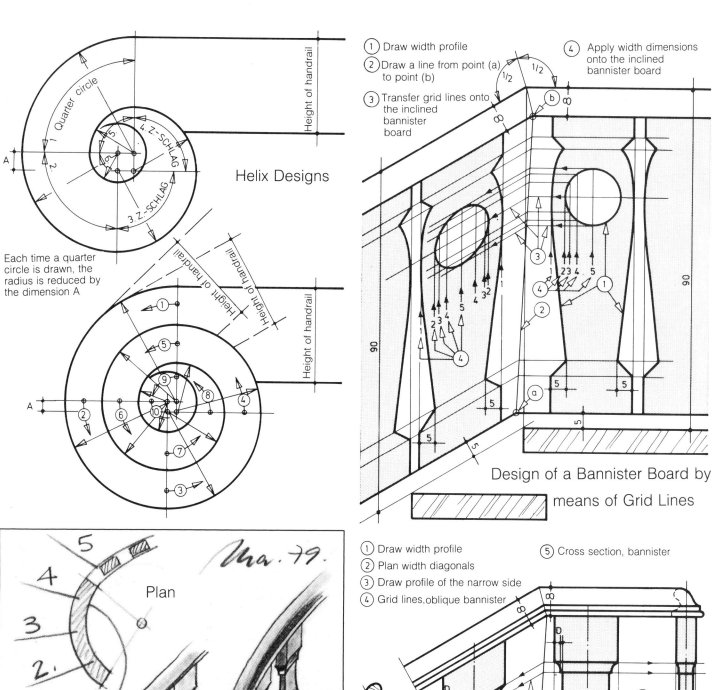

Helix Designs

1 Quarter circle

4. Z-SCHLAG

3. Z-SCHLAG

Height of handrail

Each time a quarter circle is drawn, the radius is reduced by the dimension A

Height of handrail

Height of handrail

Height of handrail

① Draw width profile

② Draw a line from point (a) to point (b)

③ Transfer grid lines onto the inclined bannister board

④ Apply width dimensions onto the inclined bannister board

Design of a Bannister Board by means of Grid Lines

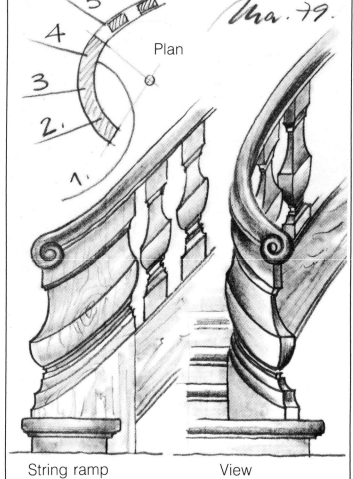

5

4

3

2.

1.

Plan

Ma. 79.

String ramp

View

① Draw width profile

② Plan width diagonals

③ Draw profile of the narrow side

④ Grid lines, oblique bannister

⑤ Cross section, bannister

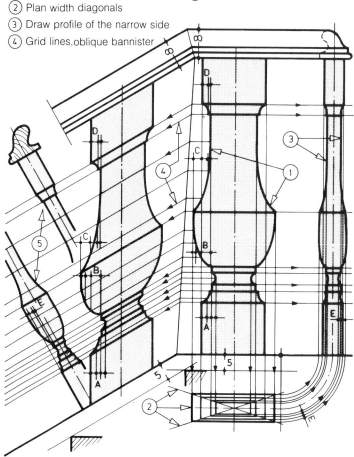

Design of Balustrade/Handrail, Arranged at Right Angles, by means of Grid Lines

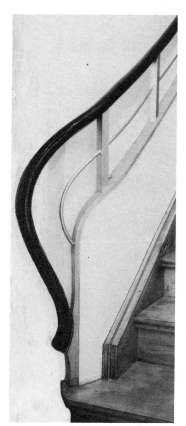

Design of stair-railing starting posts

Quarter-turn staircases without starting posts (stability provided by bannister boards placed at an angle and attached by doweling).

Balustrades

This staircase leads upward around a concrete cylinder. The steps are placed onto brackets, which pass through the concrete wall and are fastened from the inside by wedges. The same fastening methods have been used for the railing boards.

Inside the concrete cylinder, a spiral staircase leads down to a lower floor.

3.13.4. Wooden stair railings with provisions for flower boxes

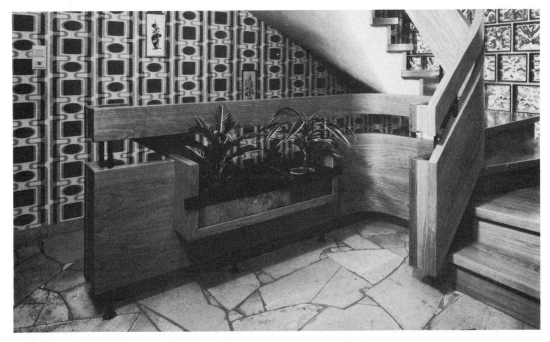

3.13.5. Stair railings with glass panels

In order to prevent accidents, panels used in stair railings should consist solely of wired glass, shatter-proof glass, or acrylic glass of appropriate thickness.

Uncovered glass edges should be ground or polished, and the corners should be beveled. Sheets of glass are available in clear, white, colored, or textured versions. Shatter-proof glass (e.g., Sekurit or Sigla) are made to measure by the supplier. Acrylic glass can be made to measure relatively easily, using specialized tools. When heat is applied, acrylic glass can be bent over formers.

Fastening:

Glass fitted into grooves should be placed onto two wooden blocks on their seating edge and surrounded with elastic material, which can offset minor movements in the fastenings.

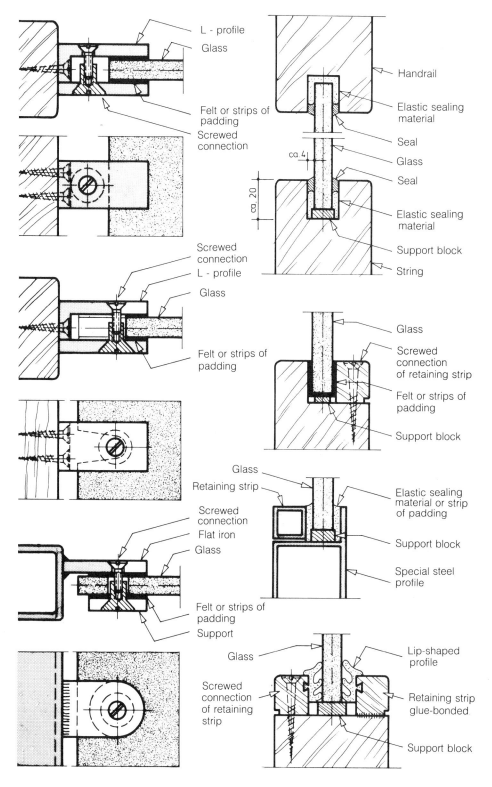

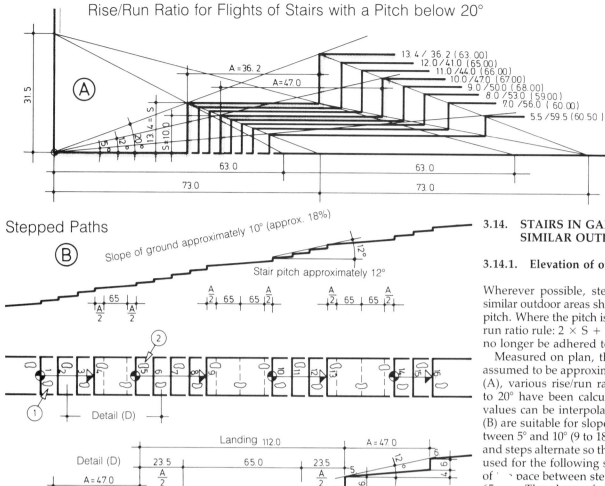

Rise/Run Ratio for Flights of Stairs with a Pitch below 20°

(A)

31.5

A = 36.2
A = 47.0

13.4 = S
S = 10.0

5° 12° 20°

13.4 / 36.2 (63.00)
12.0 / 41.0 (65.00)
11.0 / 44.0 (66.00)
10.0 / 47.0 (67.00)
9.0 / 50.0 (68.00)
8.0 / 53.0 (59.00)
7.0 / 56.0 (60.00)
5.5 / 59.5 (60.50)

63.0   63.0

73.0   73.0

Stepped Paths

(B)

Slope of ground approximately 10° (approx. 18%)

12°

Stair pitch approximately 12°

$\frac{A}{2}$ 65 $\frac{A}{2}$ $\frac{A}{2}$ 65 65 $\frac{A}{2}$ $\frac{A}{2}$ 65 $\frac{A}{2}$

(2)

Detail (D)

(1)

Landing 112.0   A = 47.0

Detail (D)   23.5 $\frac{A}{2}$ 65.0 23.5 $\frac{A}{2}$

A = 47.0

S = 10

5° 5° 12°

Stair pitch approximately 12°

Landing   5° (ca. 9 %)

3.14. STAIRS IN GARDENS AND SIMILAR OUTDOOR AREAS

3.14.1. Elevation of outdoor stairs

Wherever possible, steps in gardens and similar outdoor areas should have a shallow pitch. Where the pitch is below 20°, the rise/run ratio rule: $2 \times S + 1 \times A = 63$ cm can no longer be adhered to.

Measured on plan, the standard pace is assumed to be approximately 73 cm. In fig. (A), various rise/run ratios for pitches up to 20° have been calculated. Intermediate values can be interpolated. Stepped paths (B) are suitable for slopes in territory of between 5° and 10° (9 to 18 percent). Landings and steps alternate so that alternate feet are used for the following step (2). The length of ˙ ˙ pace between steps is assumed to be 65 cm. The slope of steps and landings should not exceed 5°. On flights of stairs where ice may form, a slope of 1° to 2° (2 to 3 percent) should be provided.

Flights of Stairs

(C)

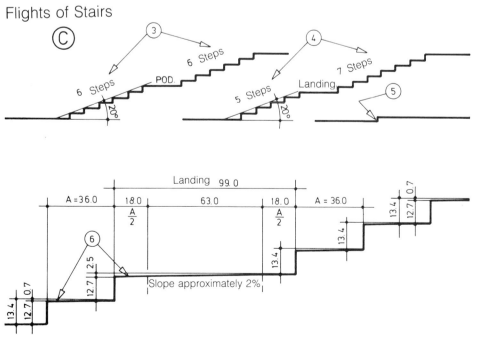

(3)

6 Steps
6 Steps   POD.
20°

(4)

7 Steps
5 Steps   Landing
20°

(5)

Where the ground slopes at a rate of 10° to 20°, steps should be built (C). Landings should be provided after 5 to 10 steps. The individual flights of stairs between landings should be of different length; i.e., not two flights of six steps each (3) but rather one flight of seven and a second of five steps (4). Individual steps can easily be missed and must therefore be regarded as hazardous (5).

The treads of outdoor stairs should have a slope of 1 to 2 percent so that rainwater can run off (6).

Landing 99.0

A = 36.0   18.0 $\frac{A}{2}$ 63.0 18.0 $\frac{A}{2}$ A = 36.0

13.4 12.7 0.7

(6)

13.4   13.4

12.7 2.5

Slope approximately 2%

13.4 12.7 0.7

3.14.2. Technical hints on the construction of stairs in gardens and other outdoor areas

When designing flights of stairs, the following rules must be borne in mind:

1. Treads and landings must be arranged so that rainwater can run off at the front and one side (slope of approximately 2 percent).
2. Appropriate drainage must be provided for rainwater to run off below stairs.
3. The construction of the stair foundations must be in keeping with climatic conditions, the nature of the subsoil, and the size of the flight of stairs.
4. In the case of large flights of stairs, the foundation must be frostproof (80 cm on all sides).

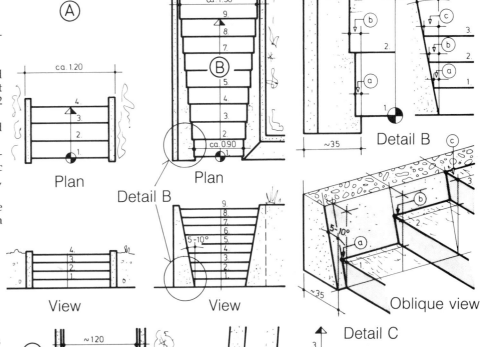

Stairs between walls

For stairs with up to four steps, the walls on either side may be vertical (3.14.2.A); above four steps the walls should be tapered toward the top (B). This taper should be between 5° and 10° (1). In example (B), the top edges of the walls are parallel (approximately 1.50 m). The bottom step in this example is only about 90 cm wide.

The solution chosen for fig. (C) is more favorable. The length of the steps is a uniform 1.20 m, while the cross section of the wall is tapered (approx. 1.65 m to 1.20 m), which makes this flight of stairs more pleasing.

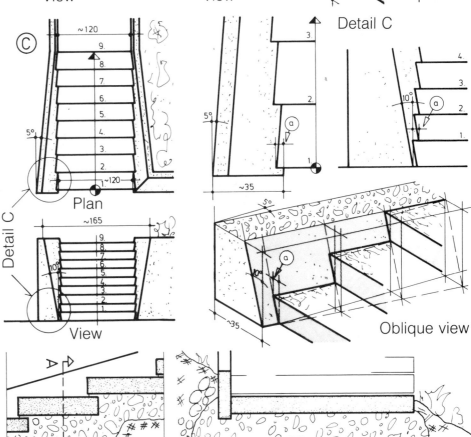

Inclined stairs along a slope

Where the ground forms a shallow slope, vertical edging stones (1) cemented into the structure are adequate. The foundation must be designed so that sliding of the stairs is prevented (2). Drainage must be provided on the hillside (3).

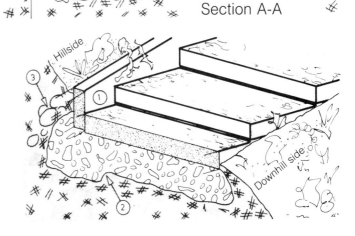

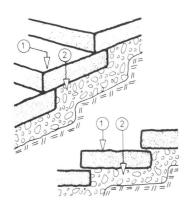

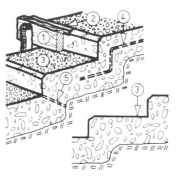

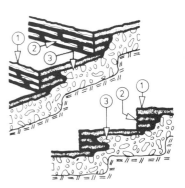

Stairs made from in-situ cast concrete
Formwork to the top of the step (1), Triangular fillet (2), Surface smooth or washed (3) (washed concrete). Reinforcement by reinforcing mats (4), Expansion joint after three to four steps (5).

Solid rectangular step (rough stone)
Solid rectangular step (1), Concrete foundation (2).

Stairs consisting of slab steps
Slab step (1), Riser (2), Undercutting (3), Concrete foundation (4)

Slate steps
Thin slate or natural stone slabs (1),
Front and edge sawn (2),
Concrete foundation (3).

Sources

Ornamental Ironwork Railings and Railing Components:
Firma BIMA, Hans Bieg, Mannheim, pp. 84, 85
Firma Emil Schneider GmbH, Werkstätte für Schmiedekunst Essingen, pp. 86, 87 top, 88
Steel Staircases and Steel Handrailings:
Firma Spreng and Co., Treppenbau, Schwäbisch Hall, pp. 89, 90, 91

   With the exception of the ornamental iron and steel handrailings, all the wooden stairs and stair railings shown in this book were made in the author's workshops.

   Assistants in the author's sutdio and workshops:
Christina Bischoff, Karin Menzel, Albert Wagner, Emil Tischer, Ernst Rosensprung, Johann Schmied, Eugen Vaas, Georg Salat, Alexander Schumann, Ivon Rogić, Alois Beck, Heinrich Binanzer, Wolfgang Nitschke.
Sculptors:
Hans Scheble, Ellwangen
Sculpture Design: Willibald Mannes, Oberkochen
Turnery Work:
Georg Liebert, Heidenheim/Schnaitheim
Photographic Sources:
Works Photographs BIMA, pp. 84, 85
Works Photographs Willibald Mannes, Photographer: Klaus Werner, pp. 39, 47, 49, 50, 51, 54, 55, 62, 63, 76, 77, 78, 79, 92, 93, 95, 96, 97, 98, 99, 100, 101, 104, 105, 106
Works Photograph Emil Schneider, p. 88
Works Photographs Spreng and Co., pp. 89, 90, 91

DK 69.026:001.4 — German Standards — November 1979

Stairs
Terms

Contents

1. APPLICATION AND PURPOSE

This standard defines the usual and common terms[1] for stairs as used by the building industry[2]. As long as the meaning of a term within its context remains clear, the prefix "stair" may be omitted.

Note: For example, landing *instead of* stair landing, handrail *instead of* stair handrail, etc.

2. BASIC TERMS

2.1. Stair
Structure consisting of at least one flight of stairs.

2.2. Interfloor Staircase
Staircase leading from one floor to the next floor
- between two full floors
- between cellar and ground floor (basement staircase)
- between the top floor and attic (attic staircase)

1. The numbers in brackets behind the terms relate to the numbering in the International Standard ISO 3880/1.
2. This standard applies to other technical fields (for example, ship building, or engineering), unless specific standards for these areas contain opposing statements.

2.3. Adjustment Step
As a rule, a step between the entrance level and the first full floor (ground floor) as well as a step connecting different levels within one floor.

2.4. Necessary Stairs
Stairs that must be provided under statutory regulations (for example, provincial building regulations).

2.5. Additional Stairs
Additional stairs that may, under certain circumstances, be the main flight of stairs.

2.6. Flight of Stairs [3]
Uninterrupted sequence of at least three steps (three risers) between two levels.

Continuation pp. 2 to 7
Explanations p. 8

DK 69.032.2:69.026.1 — German Standards — October 1953 — DIN 4174

Floor Heights and Stair Rises

Dimensions in mm

Floor heights with number and height of steps in stair flights

	Double-flight staircases				Single-flight, triple-flight, and, curved staircases (number of steps as in columns b to e) or:			
	Shallow (comfortable) rise		Steep Rise		Shallow (comfortable) rise		Steep rise	
Line	Floor height	No. of steps	Height of steps	No. of steps	Height of steps	No. of steps	Height of steps	No. of steps	Height of steps
	a	b	c	d	e	f	g	h	i
1	2250	—	—	12	187.5	13	173.0	—	—
2	2500	14	178.5	—	—	15	166.6	13	192.3
3	2625	—	—	14	187.5	15	175.0	—	—
4	2750	16	171.8	14	196.4	—	—	15	183.3
5	3000	18	166.6	16	187.5	17	176.4	—	—

1. FLOOR HEIGHTS

1.1 Floor heights are measured from top of floor to top of floor. They are listed in column a.

1.2 The figures given in lines 1, 2, 4, and 5 should preferably be used. The height of 2250 mm (line 1) only applies to cellars. Additional floor heights increase at the rate of 250 mm, for example 3250 mm, 3500 mm, etc.

2. STAIR RISES[1]

2.1. Stair rises are given in columns b to i. To facilitate the use of stairs by women, children, disabled persons, and invalids, the favorable values in columns b, c, f, and g should be used where possible.

1. Standards for other stair dimensions are being prepared.

2.2. In order to simplify design, only even numbers of steps should be used in the case of double-flight stairs—i.e., flights of equal length (columns b to e).

2.3. For all standardized floor heights, the number of steps given in the table can only be obtained with equal steps for columns b, c, f, g, and d, e, h, i, if designs are chosen where the run/rise ratio can be modified within certain limits.

2.4. In the case of apartment houses, the pitch of the stairs leading up from the cellar may often not be uniform, since steps making up for the difference between entrance and ground-floor level are determined by the rise of the main staircase, while the stairs leading up from the cellar may possibly be somewhat steeper. However, the height of the latter steps should not exceed 196.4 mm (see line 4, column e).

2.7. Walking Line [21]

Imaginary line indicating the usual path of stair users.

The graphic depiction of the walking line in a plan (see DIN 1356) indicates the direction of the stairs; the point marks the front edge of the bottom step; the arrow the front edge of the top step (see illustrations 4 to 17; the arrows in the examples indicate the direction in which the stairs rise).

The walking line lies within the walking area. *Note: This path, used by people using the stairs (ISO Standard 3880/1), cannot be clearly defined. The path actually taken by the stair users depends on the width of the step, the position of the handrail, whether the user is going up or down the stairs, age, size, and physical state of the user.*

Independent of the actual path taken by users, one can assume that in the case of straight stairs, the walking line coincides with the center of the flight of stairs. Where the top or bottom step is tapered, or in the case of circular or newel staircases, the walking line may be off-center.

A standard determining the dimensions of the walking line is in preparation (Revision of DIN 18065, Part 1).

2.8. Landing [7]

Landing at the beginning or end of a flight of stairs, usually part of the floor ceiling.

2.9. Intermediate Landing [8]

Landing between two flights of stairs, arranged between floor ceilings.

2.10. Step [18]

Component of a flight of stairs whose function is the negotiation of differences in level, normally negotiable in one step.

2.11. Tread [12]

Horizontal part of the step (see fig. 1).

2.12. Tread [20]

Horizontal top surface of a step (see fig. 1).

2.13. Riser [13]

Vertical or near-vertical part of the step (see fig. 1).

2.14. Stairwell [17]

Open space enclosed by flights of stairs and landings.

2.15. Stair Enclosure [15]

Space reserved for stairs; also staircase.

2.16. Stair Opening [16]

Opening in ceilings reserved for stairs.

2.17. Stair Railing [1]

Normally vertical restraint surrounding flights of stairs and landings as protection against falling.

2.18. Handrail [5]

Building component installed as an aid for persons negotiating stairs; attached to stair railing and/or wall or newel.

2.19. String [19]

Component supporting steps and limiting them laterally.

2.20. Supporting Beam

Structural component that carries or supports steps.

2.21. Newel

Core section of a newel staircase.

3. DIMENSIONAL TERMS

Note: This section defines dimensional terms but does not give dimensional information. (For the latter, a standard is being prepared; at present DIN 18065, Part 1, issued December 1957, and executory regulations within the individual German provinces apply.)
The definitions of these terms are at the same time indications of how measurements should be taken.

3.1. Rise [12]

Vertical dimensions between the tread of the step and the tread of the following step (see fig. 1).

3.2. Going [4]

Horizontal dimension a from the front edge of a step to the front edge of the following step measured in the walking direction (see fig. 1).

3.3. Rise/Run Ratio

Ratio of rise and run s/a: this quotient indicates the pitch of a flight of stairs. The ratio of the individual measurements should be quoted: for example, 17.2/28 (in cm)

3.4. Undercutting

Horizontal dimension u indicating the amount by which the front edge of a step projects over the depth of the tread below. (Difference between depth of tread and going, see fig. 1).

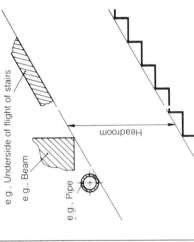

Fig. 1.

3.5. Headroom

Vertical finished dimension (measured after completion of stairs, ready for use) from the front edges of steps and above landings and the underside of structures above (see fig. 2).

3.6. Distance between Steps

In the case of slab steps, vertical finished dimension between tread and underside of the step above (see fig. 19).

3.7. Wall Distance

Distance, finished dimension, between flight of stairs or landing and the wall surface or other adjacent structural components (for example, balustrades).

3.8. Length of Flight of Stairs

Distance between the front edge of first step and the front edge of the top step, measured in the plan along the walking line.

3.9. Width of Flight of Stairs

Plan dimension of design width. In the case of laterally mortised flights, the surfaces of the unfinished walls (limiting building members) are regarded as the limit.

3.10. Utilizable Staircase Width

Clear dimension after completion (measured after completion of stairs at the handrail level) between wall surface (surface, rendering, cladding also newel) and the inner edge of the handrail or between handrails on either side.

3.11. Utilizable Landing Depth

Clear dimension measured in the plan between front edge of step and adjacent structural components (see figs. 5 to 12).

3.12. Height of Stair Railing

Vertical finished dimension measured from the front edge of the tread or the top of the landing and the upper edge of handrail or balustrade.

3.13. Width of Step (l)

Width of the smallest surrounding rectangle measured along the front end edge of the step (installed position—see fig. 3).

3.14. Depth of Step (b)

Depth of the smallest surrounding rectangle measured along the front end edge of the step (installed position—see fig. 3).

Fig. 2

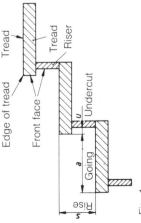

Fig. 3 Width of step (l), depth of step (b)

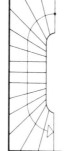

Fig. 14. Single-flight staircase with quarter turn at the top section (left staircase)

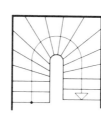

Fig. 15. Single-flight, quarter-turn angular staircase (right staircase)

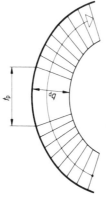

Fig. 16. Single-flight staircase with two quarter turns (left staircase)

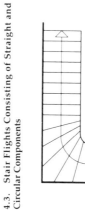

Fig. 17. Single-flight staircase with semicircular turn (right staircase)

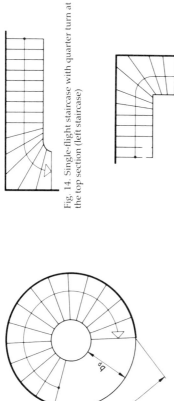

Fig. 11. Circular staircase
Staircase with stairwell
(Single-flight right staircase)

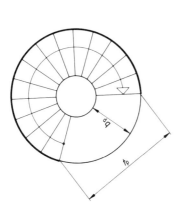

Fig. 12. Double-flight circular staircase with landing (right-hand side shows a more centered staircase)

4.3. Stair Flights Consisting of Straight and Circular Components

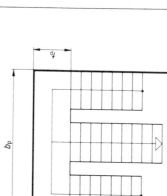

Fig. 13. Single-flight staircase with quarter turn at the starting section (right staircase)

5. TYPES OF STEPS

5.1. Types of Steps Defined by Position

5.1.1. *Bottom step*
The first (bottom) step of a flight of stairs.

5.1.2. *Top step*
The last (top) steep of the flight of stairs. Their tread (see section 2.12) forms part of the landing (see section 2.8) or half-landing (see section 2.9).

3.15. Height of Step (h)
Maximum height of individual steps in the elevation (installed position).

3.16. Thickness of Step (d)
Maximum thickness of slab steps; in the case of angular steps maximum thickness of the tread.

4. TYPES OF STAIRS [3] [4]

4.1. Stairs with Straight Flights

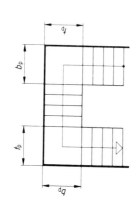

Fig. 4. Single-flight straight stair

t_P = Depth of landing
b_P = Width of landing

Fig. 5. Double-flight straight stair with intermediate landing

Fig. 6. Dogleg staircase with half-landing (right staircase)

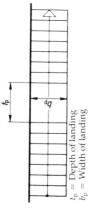

Fig. 7. Half-turn staircase with half-landing (right staircase)

Fig. 8. Triple-flight staircase, two turns, with half-landings (left staircase)

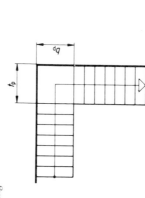

Fig. 9. Triple-flight half-turn staircase with half-landing

4.2. Curved Stair Flights

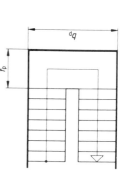

Fig. 10. Newel staircase
Stair flight with newel (single-flight left staircase)

---

3. Schematic illustration.

4. The following description is limited to the distinction between individual basic forms. A study of the history of architecture reveals highly sophisticated combinations of these basic forms, which would, however, exceed the brief outline given here.

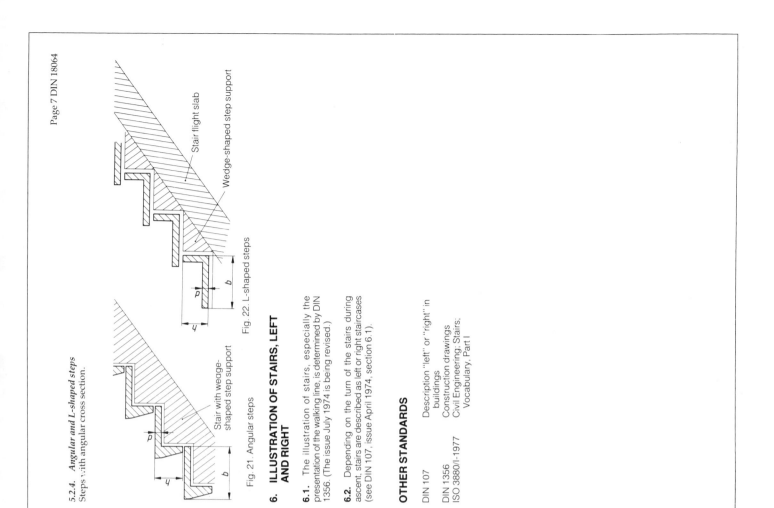

Stair flight slab

Wedge-shaped step support

Fig. 22. L-shaped steps

5.2.4. *Angular and L-shaped steps*
Steps with angular cross section.

Stair with wedge-shaped step support

Fig. 21. Angular steps

6. ILLUSTRATION OF STAIRS, LEFT AND RIGHT

6.1. The illustration of stairs, especially the presentation of the walking line, is determined by DIN 1356. (The issue July 1974 is being revised.)

6.2. Depending on the turn of the stairs during ascent, stairs are described as left or right staircases (see DIN 107, issue April 1974, section 6.1).

OTHER STANDARDS

DIN 107	Description "left" or "right" in buildings
DIN 1356	Construction drawings
ISO 3880/I-1977	Civil Engineering: Stairs; Vocabulary, Part I

5.1.3. *Leveling step*
Step between two floors with a small difference in level. More than two subsequent leveling steps form a flight of stairs.

5.2. Types of Steps by Cross Section

5.2.1. *Solid rectangular step*
Step with rectangular or near-rectangular cross section (solid or with cavity). The height of the step h is almost identical with the rise s.

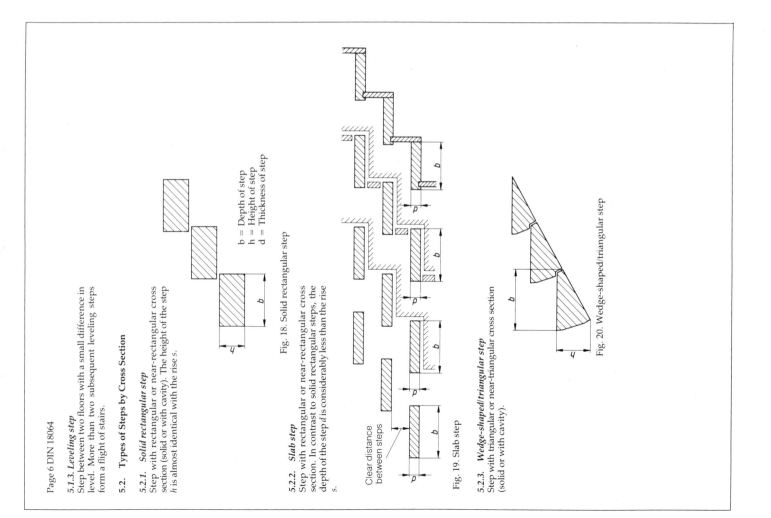

b = Depth of step
h = Height of step
d = Thickness of step

Fig. 18. Solid rectangular step

5.2.2. *Slab step*
Step with rectangular or near-rectangular cross section. In contrast to solid rectangular steps, the depth of the step d is considerably less than the rise s.

Clear distance between steps

Fig. 19. Slab step

5.2.3. *Wedge-shaped/triangular step*
Step with triangular or near-triangular cross section (solid or with cavity).

Fig. 20. Wedge-shaped/triangular step

111

Comments

It has been the object of the revised versions of DIN 18064 "Stairs: Terms, Definitions, Execution," issue August 1959, and of DIN 18065, Part I, "Stairs in Domestic Buildings, Main Dimensions," issue December 1957, to provide standards that are in line with existing statutory regulations (for example, building regulations applying in individual provinces) and existing international standards.

The present standard is the result of more than five years of work. At the publication date of this standard, listing terms and definitions, the standard on dimensions, DIN 18065 is still being revised. It is the object of DIN 18064 to provide a standardized terminology and at the same time, in section 3, clear dimensional and measuring regulations. Further standards dealing with stairs are in preparation.

The draft standard DIN 18064, Part 1 "Stairs: Terms, Definitions" was first published in May 1976. In addition to a number of new terms, it included changed terms and definitions that were not, however, adopted in practice to the required degree; consequently, the present standard does not contain major new shifts in emphasis compared with the terminology used in the previous DIN 18064, issue August 1959.

The German Standards Institution was aware of the fact that it will not be possible to combine in one standard all those terms and combinations of terms used in connection with stairs. It was therefore decided to limit the standard to those terms where a standardization of meaning with regard to uses in related standards is unavoidable.

On the basis of these considerations, the standard has been rearranged into the following sections: basic terms, dimensional terms, types of stair, and types of step.

The catalogue of basic terms contains, among others, definitions required for building inspection, such as interfloor staircase, necessary and additional stairs, walking line, etc. Terms that are difficult to define, such as main, subsidiary, and emergency stairs, have been omitted, since their exact definition appeared to be neither possible nor necessary.

The catalogue of dimensional terms contains all those dimensions that are important during planning, the obtaining of building permission, and building inspection; at the same time, the definitions indicate where and how measurements should be taken. In the case of landing width and depth, reference is made to the drawings.

The walking line of the flight of stairs is of particular significance since, in accordance with statutory regulations (for example, executory regulations for Berlin—Bau DVO, para. 12 or the relevant sections of executory regulations applying to building regulations in other German provinces), it is used for measuring the rise/run ratio, which must not change within a flight of stairs. On the other hand, innumerable circular staircases built in practice have shown that this cannot be an exact line with a constant rise/run ratio; in other words, official requirements relating to a walking line cannot be met in practice.

At best one can indicate an area within which one can talk of a "virtually constant rise/run ratio." An attempt is presently being made to find a practicable definition for inclusion in the original version of DIN 18065.

The list of types of stairs has been subdivided into stairs with straight flights, stairs with circular flights, and stairs combining straight and circular flights. Compared with the previous DIN 18064, the number of examples of stairs with straight flights has been considerably reduced and contains only the most frequently used examples. Stairs with circular flights have been subdivided into newel staircases, circular staircases, and staircases with circular sections.

The list of types of steps is, on the whole, identical with that given in the previous standard.

The international standard ISO 3880, Part 1, has been largely incorporated; the Working Group did not find it possible, however, to use it as a German standard in the form of a DIN-ISO standard without modification, since the contents of the international standard are not sufficiently comprehensive for German conditions and since special connections with existing legislation (e.g., building regulations issued by the individual German provinces and relevant executory regulations) had to be taken into account.